Handbook of Sound Studio Construction

Rooms for Recording and Listening

F. Alton Everest

Ken C. Pohlmann

New York Chicago San Francisco
Lisbon London Madrid Mexico City
Milan New Delhi San Juan
Seoul Singapore Sydney Toronto

Copyright © 2013 by The McGraw-Hill Companies, Inc. All rights reserved. Printed in the United States of America. Except as permitted under the United States Copyright Act of 1976, no part of this publication may be reproduced or distributed in any form or by any means, or stored in a data base or retrieval system, without the prior written permission of the publisher.

1 2 3 4 5 6 7 8 9 0   DOC/DOC   1 8 7 6 5 4 3 2

ISBN 978-0-07-177274-7
MHID 0-07-177274-X

Sponsoring Editor
Judy Bass

Editing Supervisor
Stephen M. Smith

Production Supervisor
Pamela A. Pelton

Acquisitions Coordinator
Bridget L. Thoreson

Project Manager
Harsimran K. Tikka,
Cenveo Publisher Services

Copy Editor
Seema Soni,
Cenveo Publisher Services

Proofreader
Alekha Jena

Art Director, Cover
Jeff Weeks

Composition
Cenveo Publisher Services

Printed and bound by RR Donnelley.

McGraw-Hill books are available at special quantity discounts to use as premiums and sales promotions, or for use in corporate training programs. To contact a representative, please e-mail us at bulksales@mcgraw-hill.com.

This book is printed on acid-free paper.

Information contained in this work has been obtained by The McGraw-Hill Companies, Inc. ("McGraw-Hill") from sources believed to be reliable. However, neither McGraw-Hill nor its authors guarantee the accuracy or completeness of any information published herein, and neither McGraw-Hill nor its authors shall be responsible for any errors, omissions, or damages arising out of use of this information. This work is published with the understanding that McGraw-Hill and its authors are supplying information but are not attempting to render engineering or other professional services. If such services are required, the assistance of an appropriate professional should be sought.

Dedicated to the memory of F. Alton Everest

About the Authors

F. Alton Everest (deceased) was a leading expert and authority in the field of acoustics. He was an emeritus member of the Acoustical Society of America, a life member of the Institute of Electrical and Electronics Engineers, and life fellow of the Society of Motion Picture and Television Engineers. He was cofounder and director of the Science Film Production division of the Moody Institute of Science, and was also section chief of the Subsea Sound Research section of the University of California.

Ken C. Pohlmann is well known as an audio educator, consultant, and author. He was director of the Music Engineering Technology program, and is professor emeritus at the University of Miami in Coral Gables. He is a fellow of the Audio Engineering Society, consultant for many audio companies and car makers, and consultant in patent-infringement litigation. He is author of numerous articles and books including *Principles of Digital Audio* (McGraw-Hill), now in its sixth edition, and coauthor of *Master Handbook of Acoustics* (McGraw-Hill), now in its fifth edition.

Contents

Preface

Imagine a cold winter morning 30,000 years ago. You emerge from a low hut and begin walking. The air is still and all is silent except for the sound of the snow crunching under your feet. You pass through a valley where you shout aloud, and the valley answers you in a series of faint calls. And when you pass a certain rock face, you know that it too returns the sound of your voice but only with one quick response.

Then you arrive at a cave mouth. You light a torch from the embers you carry in a pouch and enter the cave. Immediately you feel the relative warmth as you descend into the earth. In the darkness lit only by your small flame, foremost is the sensation of being enclosed; instead of the open sounds of the outdoors, in this place sound is strangely contained and enveloping. In tight passages, even soft sounds seem louder, while in the vast open caverns, sound repeats endlessly. You know that you are in a special place. You stop before a smooth cave wall and set down your torch. You withdraw pigments from your pouch and, singing softly in the flickering light, you begin to paint.

From our earliest days, we have marveled at the invisible presence of the sounds all around us. Whether they were sounds of animals, sounds of nature, sounds of musical instruments, or our own speaking and singing voices, we intuitively understood the importance of creating and listening to sounds. Moreover, long before science quantified the phenomenon, we have known that sound is affected by its environment. The difference between sound in the open air and sound in a cavern powerfully defines the unique nature of those places.

From these early observations, our curiosity surrounding sound evolved into a science. The Greek philosphers Pythagoras and Aristotle contemplated acoustics (our English word is derived from a Greek word) as both a musical and scientific phenomenon. Knowledge of acoustics allowed the Greeks to build open-air auditoriums, and the Roman architect and engineer Vitruvius described the acoustical properties of theaters. Galileo, Newton, Helmholtz, Rayleigh, and many others built the foundations of our modern understanding of acoustics. Places of worship, concert halls, auditoriums, recording studios, television studios, movie theaters, home theaters, hospitals, libraries, museums, art galleries, airports, and even hotels near the airport must be designed and constructed with acoustics in mind.

All of which brings us to this book, *Handbook of Sound Studio Construction*. The previous edition was authored by F. Alton Everest. He was an acknowledged authority in acoustics and a leading author in the field. During his long tenure, his many articles and books informed the design and construction of countless building projects and inspired a generation of students to choose a career in acoustics. His passing, at age 95, marked the end of an era.

While teaching acoustics at the University of Miami for many years, I often used books by Mr. Everest as required or recommended texts. I often contemplated writing an acoustics book of my own, but was well aware that excellent books such as those by Mr. Everest already dominated the market. I was thus enthusiastic, and humbled, when McGraw-Hill asked me to prepare new editions of his books.

This book, above all, strives to provide practical information on the design and construction of sound-sensitive spaces. The emphasis is on practical material that will directly assist the reader in planning a project, commercial or personal, big or small, for the recording or playback of sound. The opening chapters provide background descriptions of acoustical phenomena, but even these theoretical discussions relate to practical applications of the theory. The closing chapters present a series of design examples ranging from a simple announce booth to a recording studio. It is unlikely that any published room design will exactly fit your needs, or that you would even build it precisely as described. Rather, each design given here is a template, an example, a teaching tool that will help guide you to successful completion of your project.

If you wish to dig deeper into the study of sound, I respectfully suggest *Master Handbook of Acoustics*, another of Mr. Everest's books that I was asked to revise. It emphasizes material with more theoretical depth, and also contains additional examples of room designs. Together, these books will provide a good understanding of theoretical and practical room acoustics. With introductory books under your belt, you can widen your scope by accessing the many other acoustics books and countless magazine and journal articles available. Manufacturers of acoustical materials and treatments also publish a wealth of technical information. Finally, don't forget to consult acousticians and other professionals who have devoted their careers to the study of architectual acoustics.

Whether you are reading this book as an academic exercise or as the preface to a construction project, I welcome you to the world of room acoustics. You are following directly in the footsteps of our early ancestors, Greek and Roman architects and engineers, 19th-century scientists, designers of modern buildings, and anyone who appreciates the importance of the acoustics of spaces.

Ken C. Pohlmann
Durango, Colorado

CHAPTER 1

Introduction to Room Acoustics

Try this simple experiment: Go to an open field, and make some noise. Shout, sing, bang two rocks together, whatever. You'll note that although the sound may be loud, you can also feel it dissipate into the open air. The openness makes sounds seem empty. Now repeat the experiment indoors. You'll hear a closeness and fullness to the sound as the room surfaces return some of the sound to you. You are enveloped by the sound; it is all around you; it is more involving. The room embellishes the sound. As you explore different rooms, you'll observe that each room has a different sound quality. Simply by listening, you can tell whether a room is "dead" or "live" and whether the room sounds "dull" or "bright." Furthermore, if you listen carefully, you can hear if you are standing near a reflective wall, or far from it. You can also detect directionality and room size; for example, you can hear if your room has a high ceiling.

We conclude that rooms imprint their sonic characteristics on sounds within them. This is logical because sound emanating from a source will travel outward, strike room surfaces such as the floor, ceiling, and walls, and bounce back. The characteristics of each surface thus affect the sound that is returned to the listener. Some sound components are uniquely free of room effects. Imagine that you are sitting near a noisy machine. Some sound radiates from the machine and travels directly to your ears. Because that direct sound does not strike a room surface, it is not affected by the room. (However, its high-frequency response will be slightly reduced as it travels through air.) On the other hand, other sound from the machine strikes a room surface and returns to your ears; that sound is affected by the surface characteristics.

In any case, rooms have their own "sound" because they impose their own characteristics on audio signals contained within them. Let's think about that for a second. It's actually kind of remarkable. Sound such as music coming from headphones will sound the same everywhere. No matter what acoustical environment we are in, the headphones sound the same. That's because the room is not part of that playback signal path. But sound such as music from a loudspeaker will sound different in every acoustical environment. Every room where you set up the loudspeaker will cause the sound you hear to be different—sometimes dramatically different; that is because the room is now part of the signal path. Also, in the same room, the loudspeaker will sound different when it is placed in different locations in the room and it will sound different as you move around the room. By the same token, when you are recording a musical instrument, the sound you receive at the microphone will be different in every room and the recorded sound will sound different as the instrument or the microphone is moved. Clearly, acoustical environments such as rooms are a big deal.

We are familiar with the idea of an electrical signal passing through a black box that changes the signal passing through it. We can imagine that the box has knobs and buttons that let us manipulate the changes. A room operates the same way on an acoustic signal, and we can just as surely manipulate the changes it imposes. Instead of knobs and buttons, we use room size and geometry, glass fiber, drywall, carpet, ceiling tile, and other common and specialized construction materials to tune the room to the desired result. Getting the desired result is what this book is all about.

It's also worth noting that with a box with knobs and buttons it is easy to effect changes and vary them at will. A room's acoustical characteristics, on the other hand, are more permanent. And, we won't fully know how the room sounds until after it is built. Clearly, it's important to be able to predict the room's response with reasonable accuracy when it still only exists as a set of blueprints. Doing this requires knowledge, experience, and some mathematical tools. Finally, all the design work in the world won't yield the desired result unless the construction is done right. Relatively small differences in construction technique can spell the difference between good and bad acoustical performance.

Importance of Room Acoustics

Why do we care about room acoustics? Despite the fact that we spend most of our lives inside rooms, most people pay little attenuation to the "sound" of those rooms. That, of course, is a mistake. The sound of some rooms directly determines the success or failure of the room. For example, a concert hall that is visually stunning, but makes an orchestra sound lousy, is a disaster. Likewise a recording studio control room that causes mixing engineers to consistently turn out recordings with a deficient bass response, or a hotel room that denies a good night's sleep because of the busy interstate nearby will not be in business very long. At home, we might obsess over a room's architectural flair and its décor, but its sound is often ignored. While you might easily hear a conversation with the person next to you, you might also be subjected to noises from the kitchen, laundry room and upstairs plumbing. And imagine a lavish home theater that makes your expensive components sound mediocre.

Admittedly, for many rooms, the acoustical performance is not a top priority. But in other rooms, acoustical performance is paramount. These acoustically sensitive spaces require careful planning and construction, as well as expert evaluation. Also, clearly, acoustical performance carries a price. The additional design work, special construction materials, and skilled labor that bring premium performance all also add to the bottom-line cost.

Room acoustics can make a profound impression on a listener in ways that are hard to quantify. For example, imagine that you enter a concert hall foyer that is heavily damped. There is almost no reverberation and even if the space is physically fairly large, it seems closed and small. Now you walk into the concert hall itself and the sudden change in acoustics is evident. You hear that even small sounds are magnified and linger with a luxuriously long reverberation time. The concert hall acoustics immediately gives you a sense of immensity and a feeling of anticipation. In other words, in some rooms, the acoustics is very important.

Sound Outdoors

It is not a mistake to believe that sound outdoors is simply lost in the open air. Sound leaves a source and travels outward, unimpeded, as shown in Fig. 1-1(A). This free-field condition exists when sound is not reflected from a surface. Under this condition, sound from a point source radiates spherically outward and its sound-pressure level theoretically decreases by 6 dB with each doubling of distance from the source. For example, if the sound-pressure level of a source measures 80 dB at 10 ft, it will theoretically measure 74 dB at 20 ft. Note that this refers to sound pressure; sound intensity behaves differently. This exact theoretical result is rarely encountered in practice, but it is a handy rule for estimating sound changes with distance. Also, a point source is mainly a theoretically concept but easily approximated in practice; a source can be considered as a point if its largest dimension is small compared to the distance from it. For example, a source that is 1 ft across will act as point when measured from 5 ft or further.

Sound from a continuous line of vehicle traffic behaves somewhat differently than a point source. We assume that the sound radiates outward cylindrically (not spherically); thus its sound-pressure level decreases only 3 dB for every doubling of distance from the source. (In a continuous line, the traffic point sources reinforce each other.) This may explain why the traffic sound of the highway near your home is so annoying.

We usually, and correctly, visualize free-field conditions in an open space, but a free-field condition can also exist in an anechoic (without echoes) chamber, a specially built room with sufficient absorption to effectively absorb all energy from the source. In practice, anechoic chambers cannot quite accomplish this at low frequencies.

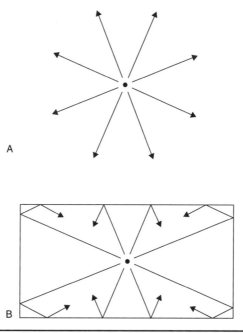

FIGURE 1-1 Sound energy radiated from a source will travel outward in different directions. (A) Outdoors, sound energy continues unimpeded. (B) Indoors, some sound energy reflects from the surfaces; this forms the basis of the room's acoustics.

Sound Indoors

Most of the acoustical properties of a room are a direct result of the effects caused by the surfaces of the room. Sound energy radiated from a source will travel outward in different directions. Some sound energy returns from the surfaces in reflected patterns as shown in Fig. 1-1(B). In some cases, sound is reflected in a more complex way known as diffusion. In either case, the returned energy comes together in a complicated way to form the sound field of the room. From a perceptual standpoint, the sound field comprises the intricate, detailed fluctuations of sound pressure at the ears of a listener. Such fluctuations test the limits of the human ear's sensitivity to sound intensity, pitch, and timbre. In addition, some sound will be absorbed by the surfaces of a room. The more energy that is absorbed by the room surfaces, the lower the sound level in the room.

Direct Sound and Indirect Sound

As noted, sound behaves differently depending on whether you are outdoors or indoors. A closer look at sound indoors shows that it exhibits both outdoor and indoor characteristics, depending on your distance from the sound source. A near-field condition exists very close to the source; sources cannot be modeled as point sources; sound decreases 12 dB for every doubling of distance. This near field should not be confused with near-field or close-field monitoring.

Slightly farther from the source (perhaps 5 ft away in a small room) the sound-pressure level decreases by about 6 dB in a free-field condition; this is direct sound. Farther than that, the sound-pressure level remains constant at any distance away; this is indirect sound. The sound-pressure level is constant because the reverberant field is constant everywhere in the room. There is a transition region between the direct and reverberant sound fields. The more absorptive the room, the lower the level of the indirect reverberant sound field. Thus, in absorptive rooms, the free-field condition extends somewhat farther from the source. In the limiting case, an anechoic chamber, the only sound is direct sound so the free-field condition extends throughout the room. The relationship between direct and indirect sound in a room is shown in Fig. 1-2.

Small-Room Acoustics

As we will see in later chapters, in many ways, small rooms pose greater acoustical challenges than large rooms. This is because in small rooms, the wavelength of sound can be similar to, or longer than, the room's dimensions. This promotes a modal response, standing waves, and a lack of diffusion. These effects create anomalies in the room's low-frequency response (for example, below 300 Hz) and other problems. In contrast, large rooms have more diffuse sound and the low-frequency response is not dominated by modes; therefore, the frequency response can be flatter. Small rooms also have relatively less absorption and thus a shorter reverberation time than may be desirable. This book deals primarily with the acoustics of small rooms. The special problems of small dimensions will appear in many of the discussions of these room designs.

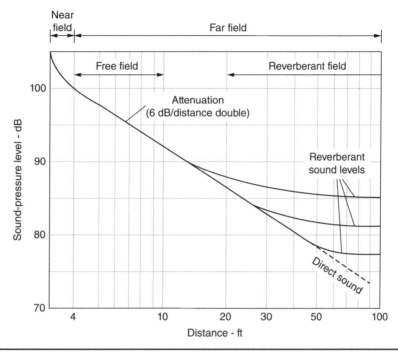

FIGURE 1-2 In a room, fairly close to the source, the sound-pressure level in the free field (direct sound) decreases by 6 dB for every doubling of distance. Farther from the source, the sound-pressure level becomes constant (indirect sound); the transition between these two zones and the level of the reverberant energy depends on the amount of absorption in the room.

Room Isolation

Let's try another experiment. While indoors, for a few quiet minutes, just listen. Can you hear traffic on the road outside, or aircraft passing overhead? Do you hear interior conversations, footstep or plumbing noise, air conditioners, or fans? Clearly, even if a room has "good acoustics," it isn't worth much if other noises intrude into the space. In other words, good isolation is an important consideration. Further, if you intend to play your home theater loudly, you'll have to consider other people in your house or your next-door neighbors. Isolation is also important for them. Isolation also depends on the frequency of the sound. Generally low frequencies are harder to isolate than high frequencies. For example, you might strongly hear the bass content from a stereo playing down the hall, but less of the song's midrange and treble content.

Adequate isolation is critical in any acoustically sensitive space. For example, recording studios, concert halls, libraries, and home bedrooms require isolation. Conversely, isolation is usually less critical in spaces such as retail stores, restaurants, gymnasiums, and home kitchens. The task of achieving isolation begins with the building's blueprints. Good isolation across the full audio frequency range usually demands heavy walls, decoupled noise sources, and other specialty architectural features. Other features might include floating floors and a specially designed heating, ventilating, air-conditioning (HVAC) system. In addition, good isolation demands attention to detail;

for example, sound leaks between rooms must be eliminated and any accidental couplings in a decoupled element must be prevented. In short, good isolation is hard to obtain. Furthermore, as noted below, good isolation should begin in the blueprints; adding considerable isolation to an existing structure is very difficult and sometimes impossible and is usually quite costly.

Room Treatment

To the average person, if they notice it at all, a room is simply an assemblage of building materials. They see a tile floor, rough stucco walls, arched ceiling, heavy wooden doors, and large glass windows. Unless the person is an architect, the room's paint scheme and furnishings might make a greater impression than the room itself.

To an acoustician, a room is a matrix of sound processing devices. To greater or lesser degrees, as shown in Fig. 1-3, every partition or barrier in a room will reflect, absorb, and transmit sound that strikes it. Reflected sound will continue to play a role, while absorbed sound will disappear. In addition, some elements will diffuse sound that strikes it; instead of a simple bounced reflection, sound is returned over a range of angles. The balancing of these phenomena (reflection, absorption, and diffusion) is key to a room's acoustical treatment. Some sound that strikes a barrier such as a wall may be transmitted through that barrier to the adjoining room. The sound that is transmitted always has less energy than the original sound because of attenuation provided by the barrier. Because of this attenuation, a barrier can provide sound isolation.

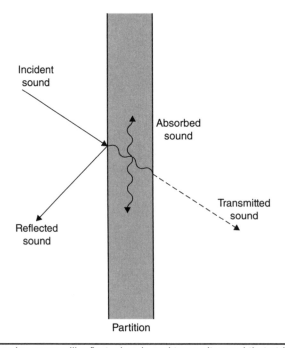

Figure 1-3 Partitions in a room will reflect, absorb, and transmit sound that strikes them. Depending on degree, a partition may be considered as reflective or absorptive, and to provide good sound isolation or not.

To an acoustician, the tile floor and stucco walls would be sound reflectors, and the arched ceiling might create troublesome sound focusing. The heavy doors (and the stucco walls) might provide useful sound isolation, while the large windows might allow noise intrusion. The paint scheme is relatively unimportant (although paint can affect factors such as absorption). Because their surface areas are relatively small, furnishings are usually less important than the building materials but, for example, a stuffed sofa and chairs would add absorption to a room.

An acoustician may also view a room as an opportunity for improvement. Through room treatment, its acoustical characteristics can be adjusted to provide more suitable performance. For example, a room with tile floor and stucco walls would have a long reverberation time. If it was a media room used for teleconferencing, the long reverberation times might reduce the intelligibility of the speech that is conveyed. To improve this, absorption in the speech frequency range could be added; for example, heavy tapestries could be hung from the walls. Similarly, heavy fabric banners could be hung from the arched ceiling to lessen focusing effects.

When treating a room, it is naturally easier to add treatment rather than take it away. For example, some acousticians prefer to design rooms that are slightly too live, and then add absorption as needed. This kind of room tuning is an important part of room treatment. It is also important to note that room treatment must closely observe the frequency response of any particular problem, and use this to design the most appropriate treatment solution. For example, if a room has an unwanted low-frequency resonance, a panel absorber could be designed with peak absorption in that particular low-frequency band.

Finally, another aspect of room treatment is the spatiality of the sound. For example, a room might be designed with absorption on one end, and diffuse reflectors on the opposite end. Or, for example, a large wood panel might be hung from the ceiling and angled so that it reflects sound to another part of the room. As with other acoustical treatment, spatiality depends on the purpose of the room. For example, a recording studio would require good diffusion throughout, whereas a home theater might require much less. In fact, in the latter, too much diffusion might degrade playback imaging, that is, the ability to localize where sounds are coming from.

Human Perception of Sound

In the end, the only thing that really matters in acoustics is the human perception of it. Because measurements and numbers are so widely used to describe sound, it is important to know what the numbers mean in human terms. The decibel (dB) is a widely used measure of sound level; it is a logarithmic ratio of two parameters such as sound powers. Sound-pressure level (SPL) compares a measured sound pressure to a reference pressure and is measured in decibels.

Because the decibel is not a linear measure, its numerical values can be misleading to the uninformed. For example, a difference of a decibel is usually not perceptible. A change in SPL of 3 dB is just noticeable; a change of 5 dB is clearly noticeable; a change of 10 dB is heard as a doubling (or halving) of sound level. Combining two signals with identical frequency and phase would double the sound pressure; this is a 6-dB increase. In the real world, with practical signals, the increase would be about 3 dB.

The logarithmic nature of decibels means that simple math cannot be used. For example, adding two 50-dB sounds does not yield a 100-dB sound; rather, the new

Difference between the Two Decibel Levels Being Added	Decibels Added to the Higher Level
0 or 1	3
2 to 4	2
5 to 9	1
10 or more	0

TABLE **1-1**   Adding Decibels

SPL Difference in Decibels between All Sources Operating and Sources Not Operating	Decibels Subtracted from SPL of All Sources Operating
0	At least 10
1	7
2	4
3	3
4 or 5	2
6 or 9	1
10 or more	0

TABLE **1-2**   Subtracting Decibels

sound level is 53 dB. Any doubling of power results in an increase of 3 dB. In fact, although it may seem illogical, 0 dB + 0 dB = 3 dB.

Table 1-1 shows that when two sound sources are played together (from the same location), the resulting sound level can be no more than 3 dB higher than one source alone. When one source is 9 dB or higher in level than the other, it dominates, and adding the second source will not measurably increase overall sound level. Some other decibel mathematics to consider: 50 dB + 50 dB + 50 dB + 50 dB = 56 dB; and 50 dB + 51 dB + 52 dB + 53 dB = 58 dB.

Table 1-2 demonstrates how decibel levels can be subtracted. This subtraction would be needed to determine what the level of one source would be if it were operating independently of other sources. For example, if the sound-pressure level of a source is measured to be 80 dB and the background noise level when the source is turned off is 75 dB, the difference is 5 dB. Thus, as shown in the table, the level of the source by itself is 80 − 2 = 78 dB.

Acoustical Design

What does it mean to design the acoustics of a room? At first glance, it may seem similar to decorating a room. While putting down carpet and hanging drapes will affect a room's acoustics, there is much more involved in acoustics. For starters, the design of a room's acoustics begins well before the room is built. The acoustics is greatly influenced by the essentials of the room's architectural design. The room's volume, geometry, and dimensions all play important roles. Moreover, the structural design of the floor, ceiling, and walls is important. For example, one type of wall design (studs and gypsum board) may absorb bass frequencies while another (masonry) does not. The room's design will also greatly influence how quiet the room will be even if it is noisy outside.

Beyond design criteria, the quality of the physical construction itself is vital. For example, even if it is well designed, a poorly constructed wall could let sound penetrate into the room. Skilled carpenters and careful interim inspection are just as important as well-conceived blueprints. Following structural construction, acoustics is further affected by the way in which the interior surface areas are treated. Different types of absorbers, reflectors, and diffusers, and their locations in the room will determine the "sound" of the room. Acoustics is also affected by the type of furnishings in the room, and even by the number of people in the room.

Clearly, in many cases, new construction is not possible, so an existing room is used. Room treatment and furnishings are thus the only tools available and depending on the initial room conditions, the ultimate performance of the room may be compromised. For example, an existing room may have inadequate isolation from outside noise, so even if the treatment is successful, the room may be plagued by noise from outside. It is also worth noting that any acoustical design is limited by the space that is available, and the budget. For example, a bass trap might be the best remedy for a particular low-frequency resonance problem, but the room might be too small to permit its installation. Likewise, sound isolation is always desirable in acoustically sensitive spaces, but it is also the most costly aspect of most projects.

It is also important to remember that different types of rooms will demand very different acoustical designs. Some rooms (such as recording studios) are used for sound production while others (such as concert halls and home theaters) are used for sound reproduction. Other rooms (such as classrooms, auditoriums, sports arenas, places of worship, hotels, airports, and restaurants) have their own unique acoustical requirements. Adding to the complexity is the fact that in some spaces the intelligibility of spoken words is the most important requirement, while in others the enjoyment of music is more important. Some rooms must accommodate both. In any case, the acoustical design of a room must be tailored to fit its specific use.

Some acoustics projects focus exclusively on noise control. For example, a room's acoustic performance may be compromised by an air-conditioning system in the building that introduces low-frequency vibrations, as well as noise at the supply vents. Solutions to the problem could be placed at the air conditioner, along the ducts, and at the vents. For example, to reduce vibrations, the air conditioner could be placed on resilient mounts, and noise could be reduced by placing silencers in the ducts and by using large-opening vents. The vibration problem is an example of structureborne noise, and the duct noise is an example of airborne noise. A complete acoustical design studies potential problems such as these and takes steps to minimize or eliminate them.

Acoustical Design Procedure

Although each project is different, the general procedure for preparing an acoustical design is usually the same. The acoustician, professional or not, usually begins with the blueprints or other drawings of the room; it might be new construction or an existing room. The acoustician must determine exactly how the room will be used; this is often difficult because many diverse demands may be made on a room. Because noise intrusion is usually a concern, a noise survey of the site is often undertaken. Based on the acoustician's analysis, the goals of the acoustical design are decided. Then, based on the size and shape of the room, building materials, and other factors, a design is prepared and integrated with the overall architectural design. Because of budget, taste, or some other necessity, plans are often modified many times before construction is completed. Clearly, it is best to finalize a design before construction begins, but this is often not possible and running changes are usually inevitable.

During the design phase, blueprints form the basis of the discussion and allow the designer to literally visualize and explore ideas and to share them with others. As a building is constructed, it is the blueprints that inform and guide the builders during the construction process. Good blueprints give builders the essential information they need including building materials, fabrication methods, and dimensions. The blueprints supplement the written contract signed between client and builder and allow verification that the work was done properly or at least according to the specifications stated in the blueprints. Moreover, blueprints document the work that has been done and serve as future reference. This record is invaluable during upgrades and renovations.

After construction is completed, the room's acoustical characteristics can be objectively and subjectively evaluated. In many cases, some treatment can be adjusted to optimize results. In some cases, some parts of the construction can be left unfinished pending final evaluation. For example, a room can be preliminarily left with deficient absorption, and then absorption can be added as necessary to achieve the desired reverberation time. It is easier and cheaper to add absorption rather than take it away. Similarly, other modifications can be made to a "finished" design to further optimize the acoustics. As noted, it should be remembered that furnishings and people will also affect a room's acoustics.

Acousticians are often asked the question—what is good sound? We will explore ways to objectively and subjectively evaluate room acoustics in later chapters, but it's worth nothing that acoustical design is certainly not a "one size fits all" situation. Different rooms require very different kinds of acoustical characteristics, and even rooms used for the same purpose can have different acoustics, and still have good sound. For example, many concert halls are admired for their fine acoustics, yet they all sound distinctly different. In fact, part of the pleasure of acoustics is hearing and appreciating these differences. In addition, listeners have different tastes and come from different cultures, and thus may have very different opinions on what the optimal sound should be. On the other hand, there is usually no disagreement when it comes to poor acoustics; it is usually apparent when speech is unintelligible, when music does not sound full and rich, or when intrusive outside noises are heard.

There is no simple solution to the design problem, but ultimately it is the duty of acousticians to reconcile many different and sometimes contradictory criteria, and to make sure their room designs provide a consensus "good sound." Acoustical design is an art and a science. It has theoretical roots in physics, material science, and psychoacoustics, but many subtle aspects of acoustical design cannot be easily explained by theory. Good acousticians acquire a feel for expert design that comes from years of practical experience.

A Note on the Room Design Examples

Following the introductory chapters in this book, the remaining chapters present a number of practical room designs and design variations. These examples range from relatively simple rooms such as announce booths to complicated rooms such as control rooms. Each room example is meant to demonstrate solutions to problems that are commonly found in those kinds of rooms. For example, the small dimensions of an announce booth mean that frequency-response irregularities caused by room modes must be addressed, and control rooms require careful design so that the sound at one particular location, the mixing position, is as neutral and accurate as possible.

However, each room-example chapter also contains more general information that may be applicable to other types of rooms. For example, the chapter describing a home project studio describes a way to build a wood cover to help insulate windows against noise intrusion. Clearly, this design detail could be applied to any kind of room. In other words, if you are interested in a specific type of room, read that chapter first. But also skim the other chapters for further information that may improve your design.

In a number of room examples, proprietary acoustical products are cited in the design. These products will certainly fulfill the acoustical requirements, but other proprietary products may be used. And, in most cases, similar systems can be constructed from scratch using common building materials. In other words, the proprietary products are examples, not particular recommendations or requirements. On the other hand, when substitutions are made, be sure to check that the acoustical properties of the new materials match those given in the room examples. Or, if the properties of the materials are different, account for the effects and modify your design accordingly.

The reader will notice a large number of illustrations in this book. In the introductory chapters, many graphs are used to illustrate the concepts explained in the text. Many graphs deserve more than a quick glance; there is often a good deal of information contained in the figure, and a greater understanding can be gained by studying it. The later chapters show a number of architectural drawings for the various room design examples. These drawings use the style of construction blueprints, and as with earlier figures, can convey a good deal of information. The plans in this book are less formal than blueprints, but should similarly convey the architectural concepts both in broad scale and in construction details. Also, by studying these drawings, the reader will gain an understanding of how the design and construction of rooms are communicated through plans.

Finally, there is no better way to understand a complex job than by doing it yourself. The reader is encouraged to use the information presented here and elsewhere, and make original drawings of structures such as absorbing panels, as well as complete room designs. By practicing and experimenting with drawings, over time, it will be possible to envision and complete a finished drawing set that can serve as the basis for construction, or at least be used to communicate ideas with an architect or acoustician.

CHAPTER 2

Sound-Reflecting Materials

The reflection of sound is perhaps the most intuitively understood acoustical property. We know that when we stand outdoors in an open field and shout, the sound is immediately lost. We deduce that most of the energy travels outward into the air while some is absorbed by the ground. But when a sheer rock wall is nearby, we hear a distinct echo. Sound energy has traveled outward, struck the wall and bounced from it, and traveled back to us. This is sound reflection. It takes some time for the reflected sound to return to us, and we observe that the further away the wall, the longer the return time. With some experimentation, we might calculate the speed of sound: sound travels about 1130 ft in one second. Now consider a rock wall with thick moss growing on it. There is an echo, but it is softer. We observe that some surfaces are better sound reflectors than others. It appears that a good reflector returns almost all the sound energy to us while a poor reflector absorbs much of the energy and returns little.

Next we go indoors and repeat the experiments. The results are similar, but also more complicated. A hard surface such as a plaster wall efficiently reflects high-frequency sound while a soft surface such as a carpeted floor reflects almost none. However, we also observe that the room enclosing us provides not just one reflection, but many. A sound might reflect from one wall, then another and another. Each trip across the room takes some time (the larger the room, the longer the time), and the result is a multiplicity of reflections spread over time. Instead of a discrete echo, we hear densely spaced reflections, that is, reverberation. Logically, rooms with highly reflective surfaces provide long reverberation times while rooms with weakly reflective surfaces offer short reverberation times.

A room's reverberation time is largely determined by the choice of surface materials. The question is thus presented: what reverberation time is best? The answer depends on how the room will be used. Any musician or music lover will tell you that reverberation is welcome. For example, concert halls tend to have luxuriously long reverberation times because it makes it easier for musicians to perform as an ensemble, and audiences like the reverberant sound. But a long reverberation time in a recording studio is undesirable because excess recorded ambience would decrease isolation between recorded tracks and make it difficult for a mixing engineer to tailor the sound of each instrument. Therefore, most recording studios have shorter reverberation times.

Also, while musicians might welcome reverberation, they will also be picky about the quality of the reverberation. For example, the frequency response of the reverberation is as important as the reverberation time. The room surface materials can dramatically affect the frequency response of reverberation. For example, a room with surfaces that reflect low frequencies and absorb high frequencies will have a relatively lower-frequency reverberation. A room with reverberation with this kind of frequency response might be considered "boomy" and thus undesirable. It is also important to note that while reverberation is highly prized in a concert hall, isolated and discrete echoes are not. Any audible echo in a concert hall or a recording studio would indicate a serious flaw in the acoustical design.

It is also important to consider the timing and directionality of room reflections. For example, in concert hall designs, early reflections (those arriving at the listener soonest) must be carefully timed to provide an adequate sense of spaciousness. Some control rooms are designed so that no early reflections arrive from the front of the room, while many reflections arrive from the back of the room within a certain time period. Clearly, the reflection of sound is one of the most acutely judged qualities of any acoustical space.

Sound Wavelength and Reflections

In many ways, the phenomenon of reflection is simple to understand. We are familiar with reflections from mirrors, as well as echoes from a distant rock cliff. Light or sound bounces from a surface, returning to the source or another position. This is specular reflection. However, not all reflections are so simple.

First, reflection is wavelength dependent. Wavelength, denoted as λ, measures the literal length of a waveform, that is, one complete cycle of a sound wave. A waveform can be measured between any two corresponding points on the cycle such as peaks or where the waveform crosses the zero axis. Looked at another way, wavelength is the distance a wave travels in the time it takes to complete one cycle. Frequency, denoted as f, is the number of cycles per second (or hertz). Frequency and wavelength are related:

$$\lambda = c/f \tag{2-1}$$

where λ = wavelength, ft
　　c = speed of sound = 1130 ft/sec
　　f = frequency, Hz

From this relationship, for example, we see that a 20-Hz waveform is about 56.5-ft long, and a 20,000-Hz waveform is a little less than 3/4 in long. Depending on the problem at hand, it is proper to refer to a sound in terms of either its wavelength or its frequency.

Wavelength of sound is important because it determines how large a surface must be to reflect a sound. Sound of a certain wavelength (or frequency) will only reflect from a surface that is sufficiently large. In particular, this is shown in the expression:

$$x > 4\lambda \tag{2-2}$$

where x = surface dimension, ft
　　λ = wavelength, ft

This shows us that sound reflects from a surface if the surface length or width dimension is greater than 4 times the wavelength of the sound. For example, a 1-kHz sine wave is about 1.1-ft long; thus it will reflect from a surface with dimension of 4.4 ft. Also, clearly, signals higher in frequency than 1 kHz will reflect from this surface. It is interesting to note that reflecting panels can thus act as high-pass filters, only reflecting frequencies that are above a certain cutoff frequency. Also, note that in this kind of specular reflection, the angle of reflection equals the angle of incidence. As we will see in Chap. 4, when $x = \lambda$, sound is not reflected; instead, it is diffused.

Reflection and Room Geometry

The geometry of room surfaces greatly influences the behavior of reflections. Parallel reflective surfaces are common in many rooms, and can be acoustically problematic if sound reflects back and forth between the surfaces. These repetitive reflections, called flutter echoes, can be very audible as a series of impulses. Flutter echo can also be perceived as a pitch or timbre coloration, which degrades sound quality and speech intelligibility.

A room can be tested for flutter echoes by simply clapping hands together and listening for a ringing or fluttering high-frequency sound. (For example, flutter echo can be heard in most stairwells because of the many parallel walls.) The experiment should be repeated at different places in the room where more focused echoes may occur. More rigorously, the reverberation characteristic of the room can be plotted; an impulse is sounded and the energy decay is plotted over time until it disappears relative to the background noise level. An echo will appear as a spike in the sound decay slope; flutter echoes will appear as a series of spikes regularly spaced over time. A plot of a typical flutter echo is shown in Fig. 2-1; the periodic spikes are clearly visible.

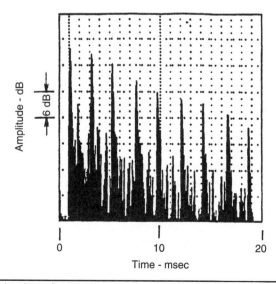

Figure 2-1   A plot showing a flutter echo as it decays over time.

Absorption is usually the easiest solution for echoes. Once the walls involved in the echo are identified, absorption can be placed on those surfaces. For example, in the case of a flutter echo, absorbing panels can be placed on one or both parallel walls. As another example, a room with a wood parquet floor and a plaster ceiling could have significant flutter echo; the floor and/or the ceiling must be treated (for example, with carpet and absorptive tiles, respectively) to eliminate the flutter echo. It is important to remember that echoes are created by specific surfaces; if absorption is placed on a surface that is not creating the echo, the echo will be unaffected. For example, in the example above, placing absorbers on the walls would not affect the floor/ceiling flutter echo.

In some cases, instead of using absorption, diffusers can be used to break up echoes. For example, diffusers would be a good choice if it is important to maintain sound energy levels in a room; adding absorption would decrease energy levels. In new construction, walls can be splayed to prevent flutter echoes; a 10:1 splay (1 ft for 10 ft of wall length) is satisfactory. Care must be taken to ensure that another (third) wall does not complete the flutter echo loop.

Reflective concave surfaces will focus sound, creating an area of higher sound level at the expense of lower level elsewhere. This is contrary to the usual need for uniform distribution throughout a room. A domed ceiling is an example of a concave surface, and a common trouble spot for acousticians. Large convex reflective surfaces, unlike concave surfaces, can provide welcome diffusion. Sound striking the convex surface reflects in many directions, distributing a broad bandwidth of sound throughout a room.

Calculating Reflections

Figure 2-2 shows some of the reflections in a rectangular room traveling from a loudspeaker to a listener. For clarity, only reflections from one loudspeaker are shown. The reflections are individually identified by letters (A–G). Through simple computations based on perfect reflections and inverse square propagation, the magnitude and delay of each reflection are estimated. In particular, the reflection level and delay can be calculated from:

$$\text{Reflection level} = 20\log\left[(\text{Direct path})/(\text{Reflected path})\right] \qquad (2\text{-}3)$$

$$\text{Reflection delay} = \left[(\text{Reflected path}) - (\text{Direct path})\right]/1130 \qquad (2\text{-}4)$$

Table 2-1 tabulates the seven reflections of Fig. 2-2, and lists values for the path lengths, levels, and delays of the reflections. We note that the direct path length is 7.1 ft. Furthermore, these values are plotted in Fig. 2-3 using the original identifying numbers. (Referring to Fig. 11-16, the lateral reflections in this room fall within the favorable region. This suggests that the lateral reflections will help provide a welcome sense of spaciousness to the room and also provide good loudspeaker imaging.)

All of these reflections are adjustable in regard to amplitude, although the delay values are fixed by the room geometry and dimensions. A given reflection can be reduced in amplitude by applying an area of absorbing or diffusing material at the point of reflection. For example, squares of absorbent on the floor, the side and front

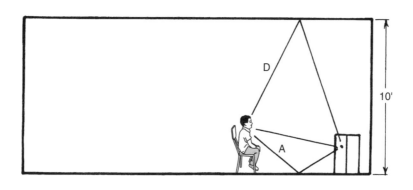

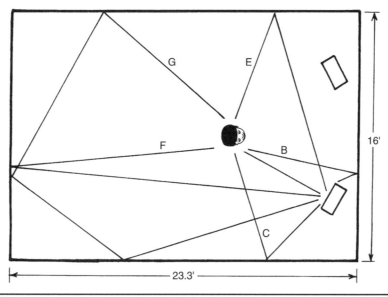

Figure 2-2 Examples of surface reflections in a listening room from a single loudspeaker to a listener.

Reflection	Reflection Path Length, ft	Reflection Path Length Minus Direct Path Length, ft	Reflection Level, dB	Reflection Delay, msec
Floor (A)	8.1	1.0	−1.1	0.9
Front wall (B)	10.2	3.1	−3.1	2.7
Near side wall (C)	13.4	6.3	−5.2	5.6
Ceiling (D)	16.4	9.3	−7.3	8.2
Far side wall (E)	20.3	13.2	−9.1	11.7
Rear wall #1 (F)	36.4	29.3	−14.2	25.9
Rear wall #2 (G)	46.8	39.7	−16.4	35.1

Table 2-1 Reflection Computations of Fig. 2-2

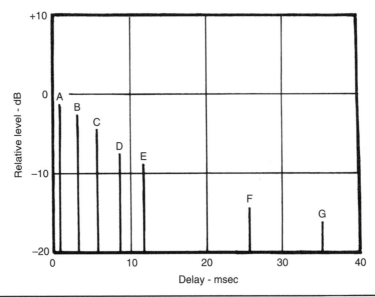

FIGURE 2-3 Plot of the listening room reflections shown in Fig. 2-2 with values tabulated in Table 2-1.

walls, and the ceiling can greatly reduce or essentially eliminate the reflections. Absorbers of different absorbing efficiency will affect the reflection amplitude. An acoustician would adjust the amplitude of the reflections to achieve the sense of spaciousness and the stereo image qualities desired. Also, the frequency response of the reflection will be varied depending on the type of absorber.

With respect to reflections, we then have a choice. Some reflections can be adjusted to provide the desired front image and degree of spaciousness, or they can be eliminated (such as the early reflections at the listening position of some control room designs).

The Precedence Effect

As noted at the outset, an echo is one of the most familiar acoustical events. At a far distance from the reflector, it takes a relatively long time for a sound to make the round trip from you to the reflector and back again. For example, at 100 ft from a reflecting surface, an echo will be heard after 177 msec (millisecond). As you move closer to the reflector, the delay is shorter. At 50 ft, the echo returns in about 88 msec. Then something unexpected happens. At a certain distance, and then as you move even closer, you cease to hear the echo.

If you have a measuring instrument, it shows that the echo is present, but you do not hear it. The "defect" occurs in your brain. The brain cannot resolve acoustical events that happen in rapid succession; the echo returns so soon after you hear the direct sound that you do not perceive the echo. There is a transition zone between perceiving and not perceiving the echo when the delay is between 50 and 80 msec, and auditory fusion is very strong at 35 msec and less.

Moreover, our brain integrates spatially separated sounds over short intervals, and tends to perceive them as coming from the location of the sound that arrives first. For example, in an auditorium, the ear and brain have the ability to gather all reflections arriving within about 35 msec after the direct sound, and combine them to give the impression that all this sound is from the direction of the original (first) source, even though reflections from other directions are involved. The sound that arrives first establishes the perceptual source location of later sounds. This is called the precedence effect, Haas effect, or law of the first wavefront. The sound energy integrated over this period is useful because it increases the apparent loudness of the direct sound without changing its perceived location.

Another issue is the loudness of the echo. The fusion zone is extended if the echo is attenuated. For example, if the echo is −3 dB below the direct sound, auditory fusion extends to about 80 msec. Room reflections are lower in level than direct sound, so in practice fusion usually extends over a longer time. However, with very long delays of 250 msec or more, the delayed sound is clearly heard as a discrete event. The fusion effect can also be overcome if the delayed sounds are increased in amplitude, but this does not occur naturally in rooms because all reflections are lower in amplitude than the original direct sound.

Reverberation-Time Equation

The reverberation time is a measure of the "liveness" of a room. Early in the development of acoustical arts and sciences, reverberation time was considered to be the most important measure of acoustical quality of a music hall. Today it is only one of many such indicators. Reverberation time is usually quoted as the time in seconds required for sound intensity in a room to decrease by 60 dB from its original level. The Sabine equation is often used to calculate reverberation time. It shows that reverberation time depends on room volume and absorption. The more the absorption in a room, the shorter the reverberation time. Likewise, the larger the room, the longer the reverberation time. This is because sound will strike the absorbing room boundaries less often. The Sabine equation is given below:

$$\text{RT}_{60} = 0.049V/A \qquad (2\text{-}5)$$

where RT_{60} = reverberation time, sec
$\quad\quad V$ = volume of room, ft^3
$\quad\quad A$ = total absorption of room, sabins

A room's total absorption A is found by summing the absorption contributed by each type of surface. This is obtained by multiplying the square-foot area S of each type of material by its respective absorption coefficient α, and summing the result to obtain total absorption ($A = S\alpha$). The absorption coefficient α describes the absorptivity of specific materials. It ranges from 0 (no absorption) to a theoretical value of 1.0 (total absorption) and varies with frequency.

For example, suppose that an area S_1 (expressed in ft^2) is covered by a material having an absorption coefficient α_1 at a certain frequency. This area contributes $(S_1)(\alpha_1)$ absorption units (in sabins) to the room. Another area S_2 has an absorption coefficient α_2, and it contributes $(S_2)(\alpha_2)$ sabins of absorption. The total absorption in the room is

$A = S_1\alpha_1 + S_2\alpha_2 + S_3\alpha_3....$ After A is obtained for the entire room, Eq. 2-5 can be used to calculate the room's reverberation time. It is worth noting that the total absorption in a room ($S\alpha$) can be increased equally well by adding a low-absorption material over a large surface area, or a high-absorption material over a smaller surface area. Also, we can see that every doubling of total absorption will cut the reverberation time in half.

The absorption coefficients of most materials vary with frequency; therefore, it is necessary to calculate total absorption at different frequencies. Moreover, any quoted reverberation time should be accompanied by an indication of frequency. For example, a reverberation time at 250 Hz might be quoted as $RT_{60/250} = 2.5$ sec. When there is no frequency designation, the reference frequency is assumed to be 500 Hz. Absorption coefficients are discussed in more detail in Chap. 3.

The Sabine equation is widely used to predict a room's reverberation time. In large spaces such as concert halls, ergodic (thoroughly diffused) conditions are approached, and the Sabine equation can be quite accurate. However, under other conditions, the equation has limitations. It is valid when reflected energy throughout a room is uniformly diffused; this is not always the case in small rooms (not ergodic). It also assumes that absorption is uniformly distributed. It also assumes that the room dimensions are similar; that is, a room does not have one dimension that is greatly different from another. In practice, the reverberation time at low frequencies may differ from that predicted by the Sabine equation; the error occurs because at low frequencies the sound is not uniform in a room. (At low frequencies, the decay of sound in the room chiefly involves the decay of a few modes.) The audible band is so wide (ten octaves) that the wavelength of sound at higher frequencies is short enough that ergodic conditions are more closely approached in small rooms. We can, therefore, depend more on the calculated high-frequency values of reverberation time.

Particularly in small rooms, reverberation time can vary considerably at different locations in the room; as we will see, this is partly due to relatively poor diffusion that occurs in small rooms. Also, the Sabine equation is more accurate for live rooms with longer reverberation times. In dead rooms where reverberation time is short, the Eyring-Norris equation will provide more accurate results.

Reverberation time describes only one part of a room's acoustics. But, it is useful to know this characteristic of a room because it can be predicted and measured with reasonable accuracy and, because it is usually consistent throughout larger rooms, a single value can be quoted. The room design examples presented later in this book contain a number of practical examples of reverberation time calculations.

Reverberation-Time Measurements

Reverberation time can be measured using an omnidirectional source and an omnidirectional microphone. The source can be sounded as an impulse. For example, electrical spark discharges, pistols firing blanks, balloon bursts, and even small cannons have been used as sources. Those signals all contain energy throughout the audible frequency spectrum. Alternatively, a steady-state source such as broadband, octave, or 1/3-octave bands of random noise can be played; the reverberation time measurement is started when the source is turned off. Sine-wave sources yield irregular decays that are difficult to analyze. The sound source is amplified and played through a loudspeaker facing into a room corner. A sound-level meter can be used to measure the sound decay; sound-level meters are discussed in Chap. 11.

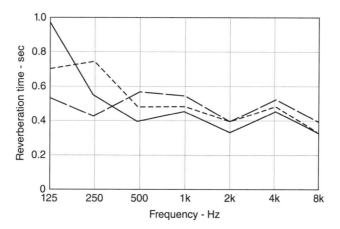

FIGURE 2-4 Measurements of reverberation time should be taken at different frequencies, and at different locations; curves for three locations are plotted in this figure. This room generally has longer reverberation times at low frequencies, as well as greater variations at low frequencies, at different locations.

The reverberation time is the time it takes the sound to decrease by a certain amount. As noted, when the sound decreases by 60 dB, the measurement is known as RT_{60}. The 60-dB figure is arbitrary, but it approximately corresponds to the time required for a loud sound to decay to inaudibility. The ear is most sensitive to the first 20 or 30 dB of decay. In practice, because of high ambient noise levels, it is often impossible to measure a sound decay over a full 60-dB range. Consider, for example, that to overcome ambient background noise, the sound source may have to be so loud as to require hearing protection for the testers. Thus it is common to measure decay over 20 or 30 dB, and then extend the curve to obtain a figure for the decay time over 60 dB. Alternatively, to increase the signal-to-noise ratio of the measurement, a band of noise can be used and the measuring device filtered to exclude all frequencies but that band.

As noted, although reverberation time is often quoted at 500 Hz, it is important to measure it at different frequencies across the audible spectrum. This can provide much useful information that a single number cannot. For example, consider the example in Fig. 2-4. The plot shows that in this room, the reverberation time is generally longer at low frequencies; this may indicate that the room has a "boomy" sound. Also, it is often useful to take measurements at different locations in a room. In small rooms, and particularly at low frequencies, there may be considerable variations in RT_{60} in different locations because of different decay rates of room modes; this variation is also shown in Fig. 2-4. To obtain a single value for RT_{60} for example, at 500 Hz, measurements at 500 Hz from different room locations can be averaged.

CHAPTER 3

Sound-Absorbing Materials and Structures

The use of absorption to reduce sound levels in a room, and to control certain reflections, is one of the most basic techniques in room treatment. Absorption directly determines a room's reverberation time, and strategic placement of absorption on room surfaces can dampen specific unwanted reflections. The optimal amount of absorption in a room depends on the room's acoustical purpose. For example, an auditorium requires relatively more absorption because too much reverberant energy will degrade speech intelligibility; a reverberation time of about 1 sec in the speech frequency range is usually recommended. However, if a room absorbs too much sound, the overall sound level in the room may be too low. Some reflection is important to maintain a satisfactory sound energy level. Interestingly, when a sound-reinforcement system is used in a large auditorium, a somewhat longer reverberation time is desirable because the amplified speech signal may sound unnatural in a highly absorptive room. A concert hall requires relatively less absorption because early reflections from the stage and a longer reverberation time are judged to benefit the sound quality of music. A concert hall with just the right amount of absorption (and other well-designed details) can yield a beautiful reverberation characteristic that is prized for both music performance and recording.

Sound absorbers are often classified as one of three types: porous materials, panel absorbers, and volume (Helmholtz) absorbers. Each of these absorbers uses the same basic mechanism to absorb sound. They convert the energy of vibration in sound waves into a small amount of heat, which is dissipated. Porous absorbers trap sound waves inside a porous material and dissipate that energy through friction. Panel and volume absorbers are similar; both dissipate sound energy through resonance.

The absorptive properties of materials are usually frequency dependent. For example, a porous absorber may be very absorptive at high frequencies, and not absorptive at low frequencies. A panel absorber may be tuned to absorb at a middle frequency region. A bass trap may absorb only very low frequencies. Furthermore, the way that materials are constructed and mounted, for example, their thickness and depth of airspace behind them, influences the frequency response over which they are absorptive. Thus a frequency-balanced room treatment often calls for a combination of different types of absorbers, using different mounting techniques.

Absorption Guidelines

The "correct" amount of absorption depends on how the room will be used, and in particular how long the room's reverberation time should be. As instructed by the Sabine reverberation equation, absorption and reverberation time are directly and inversely related. For example, if total absorption is doubled, reverberation time is halved. A long reverberation time can be problematic if speech intelligibility is paramount; absorption will decrease reverberation time and improve intelligibility. Excessive reverberation also makes it difficult to localize sound during playback. If localization is important, then added absorption may be needed.

Absorption is also used to control echoes. For example, a flutter echo comprising sound reflecting from parallel surfaces is often easily heard, and is a significant defect in a room. This echo can be reduced or eliminated by placing absorption on one or both of the parallel surfaces. As another example, a domed ceiling might focus sound at a point. This can be overcome by adding absorption to the domed surface or hanging absorptive panels underneath it.

Absorption is also used to control sound levels in a room. If the total absorption in a room is doubled, the ambient sound level decreases by 3 dB. However, this treatment is often not effective. It can be seen that if each 3-dB decrease demands that absorption be doubled, a point of diminishing returns is quickly reached. Still, a level reduction can be used to decrease noise in a room. In some cases, the reduction is a disadvantage; for example, while added absorption decreases reverberation time and improves intelligibility in a room, it also decreases level. Because of this lower level, it might be difficult for a human speaker to be heard.

Sometimes absorption can be unintentionally created, and have a negative effect on sound quality. For example, thin wooden panels on furring strips might be used for wall-covering purposes. These can act as many square feet of panel absorbers, absorbing mid- or low-frequency sound. The result might be a reverberation that is relatively too bright and harsh sounding. The solution is to eliminate the furring strips and mount the wooden panels directly on the structural surface. Likewise, too much porous absorption such as carpeting and draperies can overly reduce high-frequency sound, leaving a boomy, low-frequency reverberation.

Room Geometry and Treatment

Whenever specifying any absorption in a room, consideration must be given to the room geometry and the location of absorption in the room relative to the sound source and the listener. For example, ceiling tiles provide cost-effective absorption. But a ceiling placement is not always the best solution. In many rooms, the ceiling is relatively far from the listener. Thus a listener near a sound source will hear mainly direct sound from the source, and that sound will not be reduced by ceiling tile overhead. On the other hand, if the sound source is farther away, and the listener is hearing both direct and reflected sound, ceiling tile will help absorb sound that would otherwise be reflected from the ceiling. It can be seen that ceiling absorption is relatively most effective in a room that is long and wide with a low ceiling, and relatively less effective in a room with a small floor and tall ceiling. In other words, the room geometry must be considered when applying absorption.

In a small room with reflective surfaces and a ceiling of about 500 ft², ceiling tile might decrease indirect reverberant sound levels by about 3 dB close to the sound source, and by about 10 dB farther from the source. If absorption is also placed on the four walls, the sound level near the source is not further reduced but farther from the source the level might decrease by another 6 dB. This is perhaps the greatest level reduction that can be achieved in this room (Egan, 1988).

In many cases, when a high ceiling is available, the most efficient absorption is achieved by hanging porous absorptive panels from the ceiling. This is more effective than attaching panels to the ceiling because a greater absorptive area is exposed. Moreover, in a hanging labyrinth of panels, sound can strike multiple panels, increasing absorption.

Sound Absorption Coefficient

The sound absorption coefficient is a measure of a material's efficiency in absorbing sound energy. It is the fraction of the incident sound energy from all directions that is absorbed. The absorption coefficient, expressed as α, varies from 0 (no absorption) to a theoretical value of 1.0 (total sound absorption). For example, if 88% of the sound energy is absorbed by a given material or structure at a given frequency, its absorption coefficient would be 0.88 at that frequency. One square foot of the material would give 0.88 absorption units A, expressed in sabins. If it were a perfect absorber, such as the proverbial open window that lets all sound escape and reflects none, each square foot would give 1.0 absorption units. When using structures that allow for more than 1 ft² of material to be placed in a 1-ft² footprint, such as ceiling louvers, α can be greater than 1.0. Note that a metric sabin is defined as perfect absorption over 1 m². Absorption coefficients are dimensionless quantities; that is, units are not used.

Generally, materials with α less than 0.2 are considered to be reflecting, and materials with α greater than 0.5 are considered to be absorbing. A brick wall might have α of 0.1 at 500 Hz, while glass fiber might have α of 0.8. If materials with α less than 0.1 are added to a room, there will be little change in absorption, 0.3 would be noticeable, and 0.6 would be significant. Of course, making a significant change in a room's total absorption would require a considerably large area of absorber, not just a high α.

If the absorption coefficient of a material were determined, as it usually is, by laying the sample on the floor of a reverberation chamber, the test sound arrives from every angle and the resulting coefficient would be an average one. This is called the Sabine absorption coefficient. The Sabine coefficient, as opposed to an energy coefficient, is used in this book.

The absorption coefficient of most materials varies with frequency. Because of this, absorption coefficients are listed in tables that show the coefficients at a series of octave frequencies; for example, at 125, 250, 500, 1k, 2k, and 4k Hz. This range is far short of the audible band, but is generally sufficient for many practical applications. However, for music applications, where bass plays an important role, absorption at 31 Hz and 62.5 Hz should be considered as well.

The absorption coefficient α is a technical measure of a material's absorptivity at a given frequency; as noted, it is standardized over a 1-ft² surface. To use α in room design, we must determine how many square feet of each kind of material is in the room.

The surface area of each material is multiplied by α and summed. This yields the total room absorption A:

$$A = \Sigma\ S\alpha \qquad\qquad (3\text{-}1)$$

where A = total absorption of room, sabins
   S = surface areas, ft^2
   α = sound absorption coefficient at a given frequency, decimal equivalent

For example, a cubic room with 10-ft dimensions has surface area of 600 ft^2. Suppose all six interior surfaces are hard plaster with α of 0.1. The total room absorption is thus:

$$A = (600)(0.1) = 60 \text{ sabins}$$

From this baseline, we can see how treatment can change the room's absorption. For example, suppose carpet with α of 0.3 is placed on the 100-ft^2 floor. Also, acoustical tiles with α of 0.8 are placed on the 100-ft^2 ceiling. The plaster walls with α of 0.1 encompassing 400 ft^2 remain untreated. All values are measured at 500 Hz. Thus the new total room absorption is:

$$A = (100)(0.3) + (100)(0.8) + (400)(0.1) = 150 \text{ sabins}$$

Clearly, absorption can be added and subtracted through room treatment. Furniture choices can also influence absorptivity. For example, a heavily padded sofa will add several sabins of absorption, but metal chairs will not. It is easy to decrease a room's reverberation time by adding absorption to a highly reflective room. However, it is relatively more difficult to further decrease reverberation time in an already absorptive room. The Sabine equation for calculating reverberation time was introduced in Chap. 2.

When adjusting absorption, it is important to consider values of α at different frequencies. If the wrong materials are added, the frequency response of the room's absorption, and thus the frequency response of its reverberation, can become unbalanced. For example, adding material with α that is high at high frequencies and low at low frequencies can result in a room with relatively too much absorption at high frequencies. The resulting reverberation will contain too much low-frequency content and may sound boomy. Adding and subtracting absorption is easy. Tuning its frequency response is more difficult.

The absorption coefficients for many common building materials are listed in the Appendix. These values are taken from several references and are intended only as guidelines. In fact, tables of generic absorption coefficients should always be viewed with skepticism; there are often significant differences when measuring absorption, and specific materials of the same type often provide very different absorption values. When available, product-specific absorption coefficients provided by the manufacturer should be used. Absorption data should be provided for several octave-band frequencies. In some cases, manufacturers provide absorption data in terms of sabins; this is useful when it is difficult to calculate surface area and thus potentially imprecise to use absorption coefficients. If needed, the absorption coefficient can be calculated from the sabin value. In either case, the user must be careful to note how absorption is being specified.

Clearly, because quoted absorption coefficients are not always accurate for a given application, calculations such as reverberation time which rely on the coefficients will similarly also not always be accurate. These calculations are thus only guidelines to be used in the design phase. During and after construction, it is important to measure criteria such as reverberation time and compare the measured values to the calculated prediction. Absorption values that are calculated from a measured reverberation time will be more accurate than predicted values. Good acousticians include variability options in their designs so that final adjustments in absorption can be made; this allows for discrepancies between the predicted and actual absorption in a room.

Noise Reduction Coefficient

The noise reduction coefficient (NRC) is a single-number rating often used to specify a material's absorbency. NRC is the average of a material's four absorption coefficients α at 250, 500, 1000, and 2000 Hz:

$$\text{NRC} = (\alpha_{250} + \alpha_{500} + \alpha_{1000} + \alpha_{2000})/4 \qquad (3\text{-}2)$$

where NRC = noise reduction coefficient, decimal percent
  α = absorption coefficient, decimal percent

NRC is more informative than a single α value because it covers a number of different frequencies. However, those frequencies are only mid frequencies, so NRC is not useful for considering absorption at lower and higher frequencies. In particular, because music covers a much wider frequency range, NRC should not be relied on for these applications. On the other hand, NRC is convenient for less critical applications such as speech. It is also important to note that as with any single-number specification, NRC can sometimes be misleading. For example, two materials can have the same or similar NRC ratings, but very different absorption responses over the same frequency range.

Very generally, materials with an NRC value greater than 0.4 are considered to be very absorptive. Materials with an NRC value in the 0.4 to 0.6 range can be successfully used to provide absorption. Materials with NRC values greater than 0.8 are used when additional absorption is needed in acoustically sensitive room designs. NRC is defined in the *American Society for Testing and Materials (ATSM) Standard C423-90a Standard Test Method for Sound Absorption and Sound Absorption Coefficients by Reverberation Room Method*. NRC values are dimensionless quantities; that is, units are not used.

Standard Mounting Terminology

A material's absorptivity and the frequency response of its absorptivity can vary widely depending on how the material is mounted to a room surface. There are several methods widely used to mount absorptive materials in buildings. To account for these, manufacturers often measure absorption and publish the data using different testing standards that represent practical mounting methods. For example, laying the sample for measurement on the floor of a reverberation chamber is intended to mimic the common practice of cementing absorption directly to a wall. There are many deviations from this and standard ways of referring to different mountings. The American Society for Testing and Materials (ASTM) has standardized several methods including the ASTM designation E 795-83. Some literature may list absorption data adhering to the older Acoustical and Board Products Manufacturers Association (ABPMA). Table 3-1 lists several mounting standard designations.

ASTM Mounting Designation	Description	ABPMA Mounting Designation
Type A	Material placed directly on reflective surface	No. 4
Type B	Material cemented to plaster (gypsum) board	No. 1
Type C-20	Material with perforated or expanding facing furred out 20 mm (3/4") from reflective surface	No. 5
Type C-40	Material with perforated or expanding facing furred out 40 mm (1-1/2") from reflective surface	No. 8
Type E-405	Material spaced 405 mm (16") from reflective surface (hung ceiling tiles)	No. 7

TABLE 3-1 Common Mounting Standards for Sound-Absorbing Materials

Porous Absorbers

When sound falls on a porous surface, some of it is absorbed and some is reflected. In particular, as the sound penetrates within a porous material, the vibrating air molecules (undergoing compression and rarefaction) impart movement to the tiny fibers inside. Such movement encounters resistance as fiber rubs on fiber, and this resistance generates a small amount of heat. Thus absorbed sound energy is changed to heat energy and it appears to vanish; the amount of heat produced is so small it is unnoticeable.

There are many types of porous sound-absorbing materials. The fact that the audio band is so wide (ten octaves) results in a basic sound-absorption problem: All practical sound absorbers are frequency dependent. The determining factor in absorption is the wavelength of the sound being considered. To be a good absorber, a porous absorber must have a thickness that corresponds to a significant fraction of a wavelength. For example, at 100 Hz, the wavelength of sound is 11.3 ft (1130/100 = 11.3). It is not practical to build a porous absorber to very large dimensions; thus porous absorbers are not used at low frequencies. On the other hand, porous absorbers are most effective at higher frequencies. As noted below, the thicker the porous material, the lower the extent of the frequency response of its absorption. Alternatively, the material can be placed away from the surface behind it, ideally at one-quarter wavelength, with an airspace between the material and the surface.

Glass Fiber

In addition to its widespread use as a thermal insulator, glass fiber is widely used as a porous absorber. Glass fiber is a composite of materials such as sand, limestone, soda ash, and other components. These are the same components used to make plate glass but the material is spun into fibers. A wide variety of glass-fiber materials is available in various forms, for example, as compressed boards, semirigid boards, and loose blankets. Table 3-2 lists the 700-series glass fiber products from Owens-Corning. Type 701 (1.5 lb/ft^3) is a fluffy batt (sometimes called fuzz) useful for wall inner spaces and wherever rigidity is not required. Type 703 (3.0 lb/ft^3) is a semirigid board that cuts readily with a knife and holds its shape without support. Type 705 (6.0 lb/ft^3) is a denser and more rigid board than the 703. Glass fiber of 3-lb/ft^3 density, whether from

	Thickness		Width		Length		R-value
Type 701	1"	25 mm	24"	609 mm	48"	1219 mm†	4.2
(Density 1.5 pcf	1-1/2"	38 mm	24"	609 mm	48"	1219 mm	6.3
k-value .24)	2"	51 mm	24"	609 mm	48"	1219 mm	8.3
	2-1/2"	64 mm	24"	609 mm	48"	1219 mm	10.4
	3"	76 mm	24"	609 mm	48"	1219 mm	12.5
	3-1/2"	89 mm	24"	609 mm	48"	1219 mm	14.6
	4"	102 mm	24"	609 mm	48"	1219 mm	16.7
Type 711	1"	25 mm	24"	609 mm	48"	1219 mm†	4.0
(Density 1.7 pcf	1-1/2"	38 mm	24"	609 mm	48"	1219 mm	6.0
k-value .25)	2"	51 mm	24"	609 mm	48"	1219 mm	8.0
	2-1/2"	64 mm	24"	609 mm	48"	1219 mm	10.0
	3"	76 mm	24"	609 mm	48"	1219 mm	12.0
	4"	102 mm	24"	609 mm	48"	1219 mm	16.0
Type 703	1"	25 mm	24"	609 mm	48"	1219 mm	4.3
(Density 3.0 pcf	1-1/2"	38 mm	24"	609 mm	48"	1219 mm	6.5
k-value .23)	2"	51 mm	24"	609 mm	48"	1219 mm	8.7
Type 705	2-1/2"	64 mm	24"	609 mm	48"	1219 mm	10.9
(Density 6.0 pcf	3"	76 mm	24"	609 mm	48"	1219 mm	13.0
k-value .23)	3-1/2"	89 mm	24"	609 mm	48"	1219 mm	15.2†
	4"	102 mm	24"	609 mm	48"	1219 mm	17.4†

†Made-to-order board size.
‡Available in 703 Series Insulation only.
Made-to-order sizes are available in 1-in increments up to 48 in × 96 in. Contact your local Owens-Corning
  sales representative for minimum order quantities.

TABLE 3-2   Owens-Corning 700 Series Insulation Technical Data

Owens-Corning or from other suppliers, is widely used in acoustical treatment in diverse
places and structures, and can be considered something of a standard in the field.

Mineral Wool

Mineral wool is another widely used porous absorber. Mineral wool is a composite of
materials such as sand, basaltic rock, glass, and other components that are melted and
spun or pulled into filaments. Binder is added to form the final product. It is available
as compressed boards, semirigid boards, and loose blankets. Mineral wool is widely
used in ceiling tiles. Its acoustical properties are similar to that of glass fiber.

Density of Absorbent

One would expect density to have an appreciable effect on a material's absorption coefficient. After all, a harder surface would be expected to reflect sound more readily than

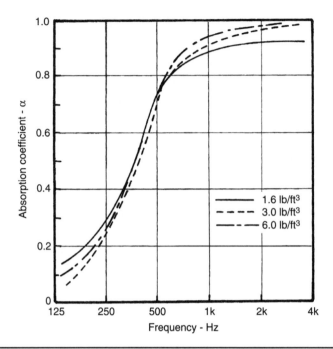

FIGURE 3-1   The effect of density on glass-fiber absorption. (*Owens-Corning*)

a softer surface. It is surprising to learn that differences are small over a normal range of density, as shown in Fig. 3-1. There is a small effect at frequencies above 500 Hz. Generally, glass fiber with a density of 3 lbs/ft³ is often used. As noted, an acoustically transparent cloth cover will not affect absorption.

Independent of the question of density, Fig. 3-1 also shows the characteristic frequency response of a porous absorber on a rigid backing. The absorber exhibits the shape of a high-pass filter with relatively little absorption at low frequencies and much greater absorption at higher frequencies. Although Fig. 3-1 is typical of a porous absorber, specific absorbers will exhibit different curves with greater or lesser absorption as well as frequency shifts in the curve. In some cases, particularly for materials such as concrete with little absorption, the curves will be much flatter with respect to frequency.

Space behind Absorbent and Thickness of Absorbent

The depth of the airspace behind a porous material, and the thickness of the material, has a great effect on the sound absorption characteristics of the material. For optimal absorption in a particular frequency region, a porous material should be placed at a distance of a quarter wavelength (maximum compression) from the reflecting surface behind the absorber or have a thickness of a quarter wavelength.

The absorption characteristic of a glass-fiber test board of 1-in thickness is shown in Fig. 3-2; the density of the board has little effect, as noted in the previous section.

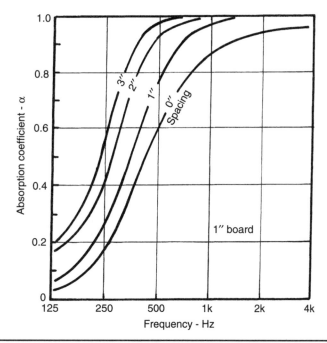

FIGURE 3-2   The effect of airspace on glass-fiber absorption. (*Owens-Corning*)

The board with no airspace (cemented directly on the wall) will be a standard of comparison. Furring a glass-fiber board out from a reflecting surface greatly increases the absorbing effect. In fact, the 1-in board furred out has the absorbing efficiency of a much thicker flush-mounted board. The achievement of greater absorption can be reduced to the question of whether the furring out is cheaper than a board of greater thickness.

Generally, absorption at low frequencies increases as the thickness of a porous absorber increases. In particular, absorption is high when the air-particle velocity is high; thus absorption is increased when the thickness of the absorber is at least a tenth of the wavelength to be absorbed, and ideally should be a quarter wavelength. Particle velocity is low close to a rigid boundary so little absorption occurs there; this explains why thin porous absorbers are ineffective. Velocity is higher farther out from the boundary so the outer surfaces of thicker absorbers can absorb longer wavelengths; this explains why thick absorbers, or absorbers spaced away from the mounting surface, are more effective. A practical limit is reached when the thickness required (determining the low-frequency limit of absorption) becomes too great. This is why porous absorbers are not usually effective for low-frequency absorption.

Figure 3-3 shows the effect of the thickness of a glass-fiber board in terms of wavelength of sound. The wavelength of sound at 250 Hz is about 4.5 ft. The quarter wavelength is about 13 in. The thickness of the 4-in glass-fiber board approaches the quarter wavelength of 250-Hz sound, and the great superiority of the absorption of the 4-in board over the 1-in board is due strictly to this fact. The conclusion is simple: to be effective over a range of frequencies, a porous absorber must be thick, or else spaced away from the boundary surface.

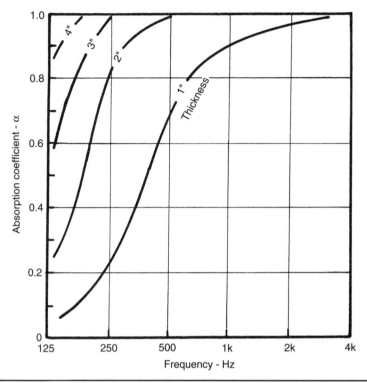

Figure 3-3 The effect of thickness of glass fiber on absorption. (*Owens-Corning*)

Handling 4-in glass-fiber boards (and even thicker) can be a problem in the acoustical treatment of a studio. Some framing structure will be needed to hold the 4-in or 6-in board and to protect it from damage. For low-frequency absorption, porous absorbers inefficiently must occupy a large volume. Thus when absorption below 125 Hz is required, attention is usually directed toward other methods of achieving it, such as resonators.

The Area Effect

As noted, the total absorption is a function of a material's sound absorption coefficient, and the amount of surface area of the material. However, total absorption is also affected by the material's distribution in the room. A number of porous panels will have greater total absorption when spaced apart from each other (for example, in a checkerboard pattern), than when placed together; this is known as the area effect (Bartel, 1981). The effect occurs because of sound diffraction at the edges of the panels, and by the additional absorption surface area provided by the exposed edges of the panels. This perimeter area is efficiently utilized because of further reflections from the hard surface near the panel edges. The area effect can also add desirable diffusion to a wall surface. On the other hand, a surface of spaced apart panels will provide slightly less total absorption than if the entire surface was covered with absorption. Obviously, the area effect provides an efficient way to minimize absorption costs.

Ceiling-Mounted Absorption

Absorption can be efficiently achieved by taking advantage of certain ways of designing ceiling-mounted absorbers. Absorbing panels can be hung from the ceiling individually or as preassembled modules. This can greatly increase the exposed absorption surface area. For example, hanging a 1-ft² panel can expose a 2-ft² area of absorption. Because the amount of absorption per square foot is much higher than the material's absorption coefficient measured over 1 ft², the effective absorption coefficient can be greater than the theoretical limit of 1.0. However, panels must be spaced apart to prevent one panel from blocking sound entry to another; in some cases, the ratio of exposed surface area to the ceiling area should not exceed 0.5.

Acoustical Tile

Acoustical tile placed on wall areas and ceilings is a cost-effective way to add absorption to a room. In addition, tile is not prone to wear and tear when placed on the ceiling. It is common to select 12 × 12-in acoustic tile with an average thickness of 1/2 in. These are cemented to the surface or placed in T-bar ceiling suspension systems. Figure 3-4 shows the spread of the data among eight different brands of 3/4-in acoustic tile. Good absorption above 500 Hz is the general rule, dropping off quickly below that frequency.

Glass-Fiber Absorber Panels

A glass-fiber blanket is an efficient absorber. It can be used to construct absorbing panels that will provide excellent wideband absorption above 125 Hz; an example is shown in Fig. 3-5. A frame, measuring perhaps 2 × 2 ft or larger, can be constructed of 1 × 6-in

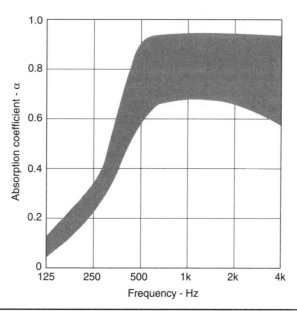

FIGURE 3-4   The general absorption characteristic of acoustical tiles.

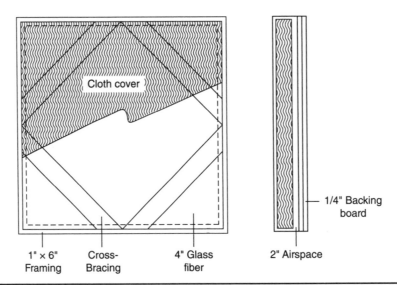

1" × 6"   Cross-   4" Glass   2" Airspace
Framing   Bracing   fiber

Figure 3-5   An example of a broadband absorber panel using glass fiber within a wood frame.

lumber, reinforced by cross-bracing. A backing board of 1/4-in plywood or Masonite adds sturdiness. The frame is packed with a 4-in thick glass-fiber blanket of 3-lb/ft³ density. Instead of a 4-in blanket, two 2-in blankets can be used. If desired, zig-zag wires can be stretched across the front of the panel to help hold the blanket in place at the front of the panel. An airspace of about 2-in behind the glass fiber helps extend absorption to lower frequencies. An expanded metal lathe can be used across the front. An open-weave cloth cover conceals and contains the glass fiber. Alternatively, manufactured cloth-covered glass-fiber panels are available. In either case, it must be verified that the cloth material is acoustically transparent; many types of cloth do not meet this important criteria.

For a given thickness, in the 125-Hz to 4-kHz range, the absorption properties of glass-fiber blankets and rigid panels are similar. High-density glass-fiber panels may become more reflective at high frequencies. For all types of glass fiber, low-frequency absorption improves with greater thickness. In some cases, porous absorbers are covered by a thin plastic or paper membrane to protect the underlying material; the membrane decreases high-frequency absorption but does not affect other absorption characteristics. When possible, this side should be placed against the wall, and not facing the incident sound.

These modules can be placed on wall surfaces. The panels are reversible; hooks can be placed on the front and back surfaces so that both the absorptive and reflective sides can be used as needed. Alternatively, the panels can be mounted on hinges so that both sides can be exposed to incident sound. As another option, a grid can be constructed and permanently mounted on the wall; absorber panels can be inserted into open spaces in the grid with either the absorptive or reflective face in front. In practice, once they are mounted, panels such as these are usually not reversed to change the room acoustics. However, the ability to reverse the panels gives some variability in mid- and high-frequency absorption when the room is initially evaluated.

Bass Traps

Bass traps can be used to provide absorption at very low frequencies. This in turn can yield a "tight" bass sound that is always desirable. There is much lore associated with bass traps. However, their design is straightforward. In particular, they can be optimized and tuned by making the depth of the trap a quarter wavelength at the desired frequency of peak absorption. In control rooms, traps are sometimes placed in the floor between the console and the observation window, or they can be set in the lower walls almost anywhere. The unused space in the rear or above the ceiling of a control room's inner shell can be converted to large traps. At least in the designs of some acousticians, better understanding of control room acoustics has reduced the need for very large traps.

The common application of the name "bass trap" to any low-frequency absorber is confusing. The term rightfully refers to a very specific kind of bass absorber. Figure 3-6 sketches the prime requisite of a bass trap, that is, a structure with a quarter-wavelength depth. The air-particle velocity is maximum at the surface of the trap, which means that glass fiber near that high-particle-velocity region is very absorptive. The sound pressure is zero at the surface. This low pressure means that the trap absorbs sound energy near the peak frequency into the trap from the nearby surrounding area.

Circumstances usually limit the practical depth of traps to 2 or 3 ft. A 3-ft trap would peak at $1130/(3 \times 4) = 94$ Hz, while the 2-ft trap would peak at 141 Hz. These frequencies are in the lower axial-mode range, which could well be handled by corner absorbers. Placing such a trap at the antinode of an axial mode would have little effect on that mode. Room modes are discussed in Chap. 5.

Panel Absorbers

Vibrating panels can be used to provide frequency-selective absorption in a room. These panel absorbers, sometimes called diaphragmatic or membrane absorbers, transform the

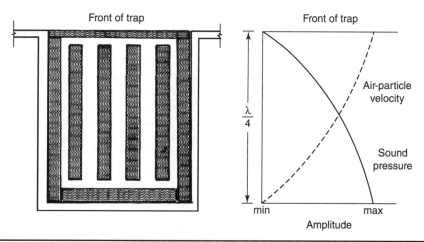

Figure 3-6 Design and layout of a bass trap, and air-particle velocity and sound pressure one-quarter wavelength from the rear of the trap.

acoustical energy of sound waves into vibrational energy, which dissipates energy as a negligible amount of heat. The airspace behind the panel acts like a spring that absorbs energy. At first glance, it may seem strange that a vibrating panel would provide absorption. However, damping inside the panel consumes more energy than the panel can radiate. Looked at in another way, in order to vibrate, a panel must draw energy away from the impinging sound wave. Panel absorbers are highly frequency dependent; they absorb sound energy that lies near their resonant frequency. The absorption characteristic of a panel absorber is typically a narrow peak; it is more difficult to obtain broadband absorption from a panel. Panel absorbers are useful for adding absorption at a narrow frequency band; however, Helmholtz resonators (described below) usually provide greater absorption.

A plywood panel, spaced out from the wall, is an example of a panel absorber. Panel absorbers are tuned to specific frequency regions and are generally used for mid- to low-frequency absorption; for example, absorption below 200 to 300 Hz is common. For this reason, panel absorbers are a useful complement to porous absorbers, which are primarily effective at higher frequencies. The frequency of resonance (peak absorption) of a panel absorber can be estimated from the expression:

$$f = \frac{170}{\sqrt{(m)(d)}} \tag{3-3}$$

where f = frequency of resonance, Hz
  m = surface density of the panel, lb/ft^2 of surface
  d = depth of airspace, in

For example, a 1/4-in plywood panel is spaced out from the wall on 2 × 4 studs (net spacing about 1/4 in). The surface density of 1/4-in plywood is 0.74 lb/ft^2. The frequency of resonance of the structure can be estimated:

$$f = \frac{170}{\sqrt{(0.74)(3.75)}}$$

$$f = 102 \, \text{Hz}$$

As shown by Eq. 3-3, the resonant (peak absorption) frequency can be lowered by using thin panels and an increased airspace. It is important to remember, however, that in practice the exact resonant frequency of a panel absorber can be different from the calculated value because of mounting methods and other factors. If a panel resonates at the wrong frequency, it will not solve an acoustical peak problem and instead could create a notch at another frequency. For that reason, some trial and error should be employed when building and installing panels. For example, a panel can be designed and installed, and then its effect should be measured in the room. Another approach to this potential inaccuracy is to install a series of panels, each with a slightly different resonant frequency. Again, care must be taken to ensure that no unwanted absorptive notches are created.

The absorption response can be manipulated (and damping increased) by placing glass fiber or similar material inside the airspace; the material should be positioned behind the panel so that it does not reduce the vibrating movement of the panel.

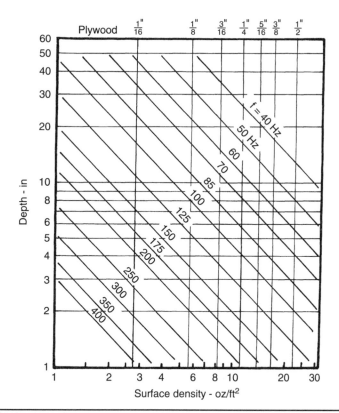

Figure 3-7 Design chart for panel resonators.

Without the material, the absorption characteristic is narrowband; when the material is added, the absorption peak is lowered, and the characteristic covers a somewhat broader band.

The minor labor of calculating the resonant frequency of a 1/4-in panel can be avoided by using the graph of Fig. 3-7, which was developed from Eq. 3-3. These panel resonators can reach a relatively sharp peak of absorption coefficient of about 0.6 at their frequency of resonance.

One method of making a practical absorber from a plywood panel is illustrated in Fig. 3-8. This 1/4-in plywood panel furred out on 2 × 4 studs resonates at 102 Hz according to our calculations above. The cavity space would be left empty if a sharp peak of absorption is desired or loosely filled with glass fiber if a flatter characteristic is desired. A 1/4-in to 1/2-in airspace between the panel and the absorbent is recommended to avoid interference with the vibration of the panel. The periphery can be isolated from the frame with strips of rubber or foam to avoid rattles. The flat structure of panel absorbers makes them ideal in rooms where space is a concern.

A panel absorber can be built into a corner as shown in Fig. 3-9. An estimate of the frequency of resonance can be made by taking the depth of the corner as the average depth. This would usually place the resonant frequency below 100 Hz, which would offer some control of axial modes. Providing absorption in a corner would efficiently affect all modes of the room since that is where they all terminate.

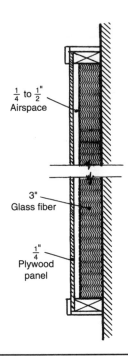

$\frac{1}{4}$ to $\frac{1}{2}$"
Airspace

3"
Glass fiber

$\frac{1}{4}$"
Plywood
panel

Figure 3-8 A wall module low-frequency panel absorber.

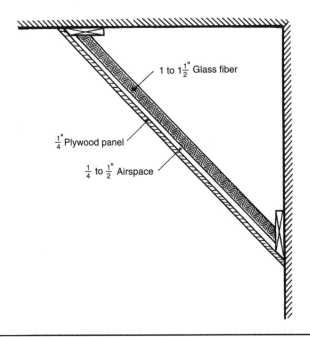

1 to $1\frac{1}{2}$" Glass fiber

$\frac{1}{4}$" Plywood panel

$\frac{1}{4}$ to $\frac{1}{2}$" Airspace

Figure 3-9 Corner modular low-frequency absorber.

Many building materials and structures such as gypsum-board walls, window glass, and wood paneling act as panel absorbers; they resonant and thus dissipate sound energy near their resonant frequency. Particularly when the materials cover a large surface area, their absorptive effect should be taken into account when estimating the total absorption in a room.

Panel absorbers only absorb low frequencies, and they reflect high frequencies. Because they are flat, the reflection is specular. Therefore, it is important to check the placement of panel absorbers to ensure that they do not contribute to a high-frequency flutter echo. If so, the panels can be relocated, angled, or covered by a porous high-frequency absorbing material.

Polycylindrical Absorbers

Polycylindrical absorbers (sometimes referred to polys) were widely used in early radio and recording studios. They served well because their acoustical characteristics are very good. They offer good low-frequency absorption, and they are dramatic in appearance. Although not as effective as more contemporary number-theory diffusers (described in Chap. 4), they also offer good diffusion over an angle of perhaps 120°.

Anyone with experience in carpentry will be curious as to how polycylindrical absorbers are constructed. Figure 3-10 shows the basic design. The space behind the cylindrical surface is segmented with bulkheads, over which a plywood skin is mounted. The top edge of each bulkhead has a strip of felt or foam to assure that the skin will not rattle against the bulkhead edge. Fabrication requires careful measurement of all elements, and accurate sawing of slots in the edge strips with a radial saw.

Absorption of Drywall Construction

Consider a typical room that is being converted into a sound-sensitive space. The walls and ceiling are ordinary drywall (gypsum) panels on stud framing. Sound in the room sets these surfaces vibrating as diaphragms cushioned by the air within them. The flexure of the surfaces encounters a resistance in the fibers, and heat is developed as a result of the vibration. This dissipation of heat means that sound energy is being absorbed. Absorption occurs at the resonant frequency of the surfaces, which is a function of the mass of the panels and depth of the airspace behind them. This effect of the wall construction may significantly contribute to the room's low-frequency absorption.

For example, drywall of 1/2-in thickness on 2 × 4-in studs spaced on 16-in centers has an absorption coefficient of 0.29 at 125 Hz; it is considerably less at higher frequencies. Thicker gypsum board and a deeper airspace yields greater absorption at

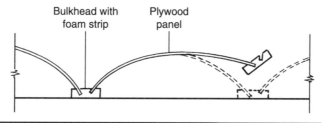

Figure 3-10   Construction details of a polycylindrical absorber.

lower frequencies. The drywall walls and ceiling are all potentially good low-frequency absorbers, and must be considered in any calculations. In some rooms with extensive drywall construction, it will be found that the low-frequency absorption is too great, and the room will have deficient bass response. This can be balanced by introducing high-frequency absorbers such as carpet; this can achieve consistent absorption across the frequency spectrum. However, the overall room absorption may then be too great.

Helmholtz Resonators

Helmholtz resonators, also commonly called volume resonators, provide absorption by reducing acoustical energy with a volume cavity; energy is consumed by air friction at the cavity mouth and by reflections inside the cavity. Helmholtz resonators are composed of a volume of air connected to the air of the room through a tube or neck. The simplest case is a bottle in which the air in the bottle is coupled to the air of the room by the air in the neck of the bottle. Blowing across the mouth of the bottle produces a tone at its natural resonant frequency. The air in the cavity is springy, and the mass of the air in the neck of the bottle reacts with this springiness to form a resonating system. Absorption is maximum at the frequency of resonance. In some practical Helmholtz resonator absorbers, holes in a perforated cover or slits between slats act as the neck of the resonator.

As with panel absorbers, the absorption frequency characteristic of Helmholtz resonators can be manipulated by placing porous absorption inside the cavity. Without the absorption, the response has a narrowband peak; when absorption is added, the response is lower and covers a broader band. In addition, the response can be broadened and damping increased by placing a porous absorber in the neck of the cavity. That portion of the sound not absorbed is reradiated in all directions. This contributes to the diffusion of sound in the room in a small way.

Helmholtz resonators appear in a variety of products. For example, Helmholtz resonators are found in specially shaped concrete blocks with the SoundBlox trade name (manufactured by Proudfoot Company). The side of the blocks facing into the room contains narrow slots that lead to the block's internal chambers. The resulting resonance yields low-frequency absorption that is built into load-bearing concrete-block walls. In some designs, the slot face of the block also has a phase grating diffusing contour. High-frequency absorption is decreased if the blocks are painted; however, paint does not affect low-frequency absorption. In addition to its absorption properties, these Helmholtz blocks, as with other concrete masonry units, provide good sound isolation. To lower shipping costs, SoundBlox can be manufactured by local licensees near the building site.

Perforated Panel Absorbers

Perforated panel absorbers are truly Helmholtz resonators, although there is little similarity in appearance between perforated panels and the bottles and cavities we typically use to describe Helmholtz resonators. In either case, the operation is identical: The air in an enclosed volume acts like a spring, and the mass of the air in the neck reacts with the springiness to form a resonating system.

Drilling many holes in a panel can approximate the necks of many volumes. Fixing the panel over a box provides a volume behind the perforated panel so that each hole has its apportioned space, completing the resonating system. The frequency at which the system resonates (when the panel is perforated with circular holes) can be calculated by the expression:

$$f = 200\sqrt{\frac{p}{(d)(t)}} \qquad (3\text{-}4)$$

where f = frequency of resonance, Hz
$\quad\quad p$ = perforation (circular hole) percentage,
$\quad\quad$ = [(hole area)/(panel area)] × 100
$\quad\quad t$ = effective hole length, in, with correction factor applied,
$\quad\quad$ = (panel thickness) + (0.8)(hole diameter), in
$\quad\quad d$ = depth of airspace, in

The perforation percentage is the percentage of hole area to panel area; it can be calculated by reference to Fig. 3-11. The perforation percentage for two types of circular-hole configurations is shown in Fig. 3-11(A) and (B), and the percentage for slat absorbers (described below) is shown in Fig. 3-11(C).

Equation 3-4 is accurate only for circular holes. The results of this equation are shown in graphical form in Fig. 3-12, which is specific for panels of 3/16-in thickness. Table 3-3 lists hole diameters and spacings to yield various peak frequencies for panel thicknesses of 1/8 in and 1/4 in for depth of airspaces of 3-1/8 in and 5-1/8 in. Figure 3-13 shows how a perforated panel resonator could be constructed in the form of an independent module.

Figures 3-14 and 3-15 are companion plots, the former for an airspace depth of 4 in with 2 in of absorbent and the latter for an airspace depth of 8 in with 4 in of absorbent. Other variables remaining fixed, note that the smaller the perforation percentage, the lower the frequency; frequencies range from 300 Hz for the shallower unit to well below 100 Hz for the deeper unit.

Particulars such as the number of holes in a unit area and size and shape of the holes influence the amount of absorption. As an alternative to circular perforations,

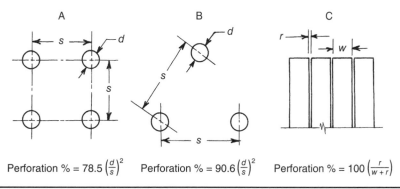

Perforation % = 78.5 $\left(\frac{d}{s}\right)^2$ Perforation % = 90.6 $\left(\frac{d}{s}\right)^2$ Perforation % = 100 $\left(\frac{r}{w+r}\right)$

Figure 3-11 Templates and equations used for calculating perforation percentage for perforated panel resonators. (A and B) The perforation percentage for two types of circular-hole configurations. (C) The perforation percentage for slat absorbers.

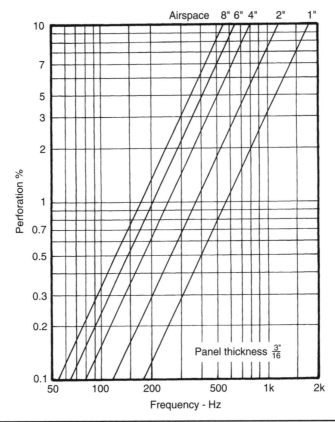

FIGURE 3-12   A design chart for perforated panel absorbers using a panel of 3/16-in thickness.

the panel may be cut through with linear saw slots, for example of 1/8-in or 3/16-in width. Panels with a higher percentage of open area are more efficient at absorbing high frequencies. Conversely, when only low-frequency absorption is required, panels with a low percentage of open area are quite satisfactory. Generally, thinner front panels are more efficient at absorbing mid and high frequencies. The front panel may be punched or pressed metal, wood or another material. The solid panel area around the holes reflects sound. Clearly, the greater the solid area relative to open holes, the greater the level of reflection.

An absorptive material can be placed behind the panel, or not. The effect of an absorber such as a glass-fiber blanket is to broaden the peak response. For maximum absorption, the blanket should be placed directly behind the perforated panel where air-particle velocity is greatest. A common volume can be used, or the volume can be compartmentalized; the latter can provide a broader absorption band. Care must be taken when painting a perforated panel with small holes so that paint does not clog the holes or the absorbing material behind them; rollers should be used, not spray.

Perforated panel absorbers can be constructed using the same framing method as described above for glass-fiber absorbers. Instead of 4-in glass fiber, a 2-in blanket is

Depth of Airspace	Hole Diameter	Panel Thickness	Perforation %	Hole Spacing	Frequency of Resonance
3-5/8"	1/8"	1/8"	0.25%	2.22"	110 Hz
			0.50	1.57	157
			0.75	1.28	192
			1.00	1.11	221
			1.25	0.991	248
			1.50	0.905	271
			2.00	0.783	313
			3.00	0.640	384
3-5/8"	1/8"	1/4"	0.25%	2.22"	89 Hz
			0.50	1.57	126
			0.75	1.28	154
			1.00	1.11	178
			1.25	0.991	199
			1.50	0.905	217
			2.00	0.783	251
			3.00	0.640	308
3-5/8"	1/4"	1/4"	0.25%	4.43"	89 Hz
			0.50	3.13	126
			0.75	2.56	154
			1.00	2.22	178
			1.25	1.98	199
			1.50	1.81	217
			2.00	1.57	251
			3.00	1.28	308
5-5/8"	1/8"	1/8"	0.25%	2.22"	89 Hz
			0.50	1.57	126
			0.75	1.28	154
			1.00	1.11	178
			1.25	0.991	199
			1.50	0.905	218
			2.00	0.783	251
			3.00	0.640	308
5-5/8"	1/8"	1/4"	0.25%	2.22"	74 Hz
			0.50	1.57	105
			0.75	1.28	128
			1.00	1.11	148
			1.25	0.991	165
			1.50	0.905	181
			2.00	0.783	209
			3.00	0.640	256
5-5/8"	1/4"	1/4"	0.25%	4.43"	63 Hz
			0.50	3.13	89
			0.75	2.56	109
			1.00	2.22	126
			1.25	1.98	141
			1.50	1.81	154
			2.00	1.57	178
			3.00	1.28	218

TABLE 3-3 Low-Frequency Panel Absorber, Perforated-Face Type

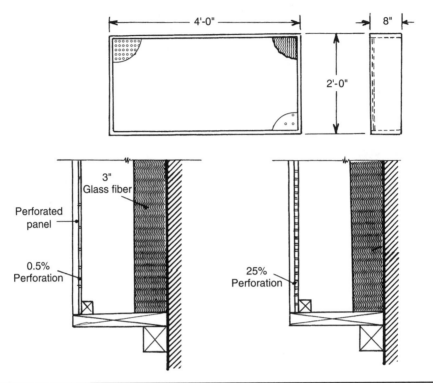

FIGURE 3-13 A wall module low-frequency panel absorber.

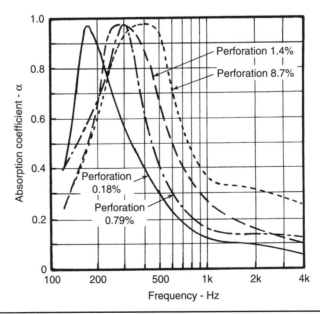

FIGURE 3-14 Absorption of perforated panel absorbers of 4-in depth. (*Mankovsky, 1971*)

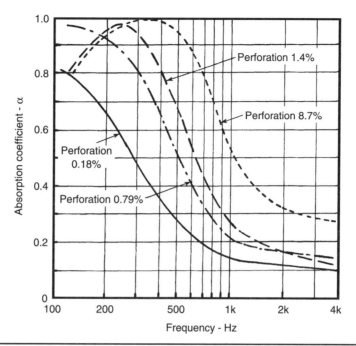

FIGURE 3-15 Absorption of perforated panel absorbers of 8-in depth. (*Mankovsky, 1971*)

used with a 4-in airspace. The front is faced with a 3/16-in plywood sheet drilled with 3/16-in diameter holes on 1-9/16-in centers. To expedite construction, a number of panels can be stacked and drilled simultaneously. Absorption of this particular panel peaks at around 125 Hz and falls off at higher frequencies.

A designer typically places the peak absorption of the low-frequency perforated unit to compensate for the low-frequency deficiencies in other absorbers to achieve consistent absorption response across the audio band. In practice, the resonant frequency can be measured by driving an amplifier and loudspeaker with an oscillator, and using a microphone or SPL meter to measure sound pressure inside the absorber placed outdoors and about 10 ft from the loudspeaker. By sweeping the oscillator frequency across the range of interest, the microphone output will show a peak at the resonant frequency; this is the frequency of maximum absorption. Clearly, the oscillator, amplifier, loudspeaker, and measuring microphone must have a flat response in the range of interest.

Slat Absorbers

Slab absorbers use narrow strips of wood placed over an airspace cavity. Although they may appear to be different, slat absorbers are full-fledged Helmholtz resonators, the hole in the "neck of the bottle" now being a short section of a slot with its proportional airspace cavity below. The mass of the air in the slot reacts with the springiness of the air in the cavity. Whether perforated holes or slats are used, they both provide low-frequency absorption. As shown [Fig. 3-11(C)], the equivalent of perforation percentage for slat absorbers is the slot area percentage. The frequency of resonance in a slat absorber can be estimated from:

$$f = 216\sqrt{\frac{p}{(d)(D)}} \qquad (3\text{-}5)$$

where f = frequency of resonance, Hz
  p = perforation (slot area) percentage
  D = depth of airspace, in
  d = thickness of slat, in

In some slat absorbers, the depth of airspace, thickness of slat, and width of slot (spacing between slats) are constant. In other designs, one or more of these dimensions are varied; this will broaden the absorption peak. Also, the absorption characteristic can be broadened by placing a glass-fiber board behind the slats and in contact with the slats.

In some designs, in addition to low-frequency absorption, slat absorbers can also provide some high-frequency diffusion. Instead of flat strips of wood, the depths of the strips are varied to create an uneven surface. An example of a slat absorber with this and other variations is shown in Fig. 3-16. With correct positioning, this type of slat absorber could also eliminate troublesome specular reflections from an otherwise flat-slat surface. Alternatively, commercially available diffuser strips can be used as slats.

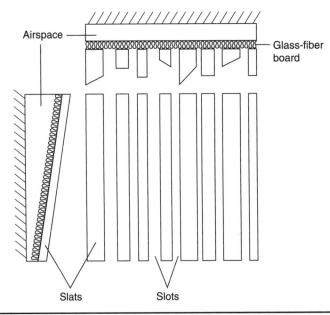

Figure 3-16 Slat absorber design showing variations in airspace depth, slat width, and slot width. These variations yield a broader absorption characteristic compared to designs with identical dimensions. Also, the slat facing is not flat; this yields some high-frequency diffusion.

Multipurpose Absorbers

In some cases, an absorber is designed to perform several tasks. This can be accomplished by combining different absorption techniques into one structure. For example, Fig. 3-17 shows an absorber mounted on a structural concrete-block wall. It uses an interior wood panel to resonate and absorb at a specific low-frequency region, as well as an outer broadband absorber. The outer pressed glass-fiber sheets and batts readily admit low-frequency sound energy that will strike the panel underneath. As with any panel absorber, the wood panel can be tuned to a specific frequency or a perforated panel could be used. Glass-fiber batt can be placed behind the wood panel. Alternatively, the wood panel could be replaced by a pressed glass-fiber sheet to further improve the broadband absorption of the entire structure.

Prefabricated Sound Absorbers

A wide variety of prefabricated sound absorbers are available for an equally wide variety of applications. While many efficient absorbers can be constructed from basic materials such as glass fiber, plywood, and 2 × 4 studs, commercially fabricated absorbers save time and may provide a more finished appearance. In addition, unlike units built from scratch, the absorption characteristics of prefabricated units are accurately known from the onset.

Ceiling tiles composed of a porous material are commonly used. Care must be taken when painting ceiling tiles and other porous absorbers; a thick coat of paint can greatly degrade absorption. For that reason, staining or tinting is preferred. In addition, because it can provide a thinner coat, spray application is better than a brush or roller. The manufacturer will recommend the best practices when modifying a porous absorber.

For wall treatment, many prefabricated units contain a porous absorbing panel placed behind an open face that protects the absorber within. The facing may be perforated metal, hard board, or metal slats. In some cases, the absorbing panel is covered by a fabric. In high-humidity conditions, the porous absorber may be covered by a thin plastic sheet. Sheets less than 1 mil thick do not greatly impede sound passing through them; however, high-frequency absorption may be decreased.

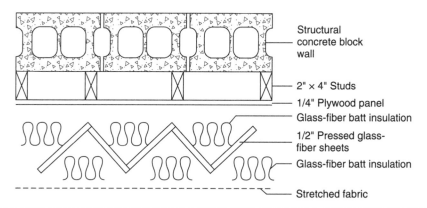

FIGURE 3-17 Multipurpose absorber construction using a panel absorber and porous broadband absorber.

Absorptive panels with a perforated metal facing are often used in outdoor applications and in environments where traditional porous absorbers would be subject to wear and tear. As with other perforated panel absorbers, the absorption frequency range is determined by the size and spacing of the holes in the front panel. If the panels are painted, care should be taken so the perforations are not clogged by paint.

Open-Cell Foams

Open-cell foams are often used as sound absorbers in sound recording and home applications; they are notable for their ease of installation. Open-cell foams are good sound absorbers, comparable to glass fiber in efficiency. The multitude of interconnected pores allows sound to deeply enter the foam, where it is trapped and dissipated as heat. Closed-cell foams have nonconnected pores that do not allow absorption of sound, and are to be avoided in acoustical applications. (However, they are effective as thermal insulators.) Closed-cell and open-cell foams are similar in appearance, but can be identified by trying to blow air through them: air goes through the open-cell type, but not the closed-cell type. It is also important to note that while open-cell foam, glass fiber and other porous materials are excellent sound absorbers, they are poor sound isolators. Sound passes through these lightweight materials; they have no "stopping power."

Sonex

Sonex molded foams (manufactured by Pinta Acoustic, Inc.) are widely used as absorptive wall coverings. Sonex is an open-cell foam (fire-retardant polyester urethane or fireproof melamine) that is sold in sheets (for example, 2 × 4 ft). Sheets can be secured to a wall surface with ordinary adhesive caulk; the caulk must be compatible with foam. Alternatively, a spray adhesive can be used. Three of the Sonex patterns, (A) SONEX-classic, (B) SONEXone, and (C) SONEXpyramids, are shown in Fig. 3-18.

The absorption characteristics of SONEXclassic (urethane, 2 lb/ft^3) are graphed in Fig. 3-19 and the characteristics of SONEXpyramids (melamine, 0.7 lb/ft^3) are graphed in Fig. 3-20. These curves can be compared to those of glass fiber in Fig. 3-3. In both cases, the Sonex has a high-pass characteristic that is similar to that of glass fiber. The quoted Sonex thickness includes the contour crests while glass fiber is solid fibrous material, making accurate comparison difficult. But, generally, because of the shape, the absorption of Sonex panels is similar to a glass-fiber panel of about half the thickness. Because of this, to increase low-frequency absorption, glass-fiber panels can be placed behind Sonex panels. As with other porous absorbers, low-frequency absorption can also be augmented by placing the absorber away from the wall with an airspace behind it.

The value of Sonex is that it is an engineered product that is ready to apply to the wall, while glass fiber requires framing and a cover. Some consumers appreciate the dramatic appearance of Sonex and use it to enhance the appearance of their studio. In other words, products such as Sonex are easy to install and have a good visual appearance, but simple absorbing materials such as glass fiber work equally well.

Are Glass Fibers Dangerous to Health?

On June 24, 1994, the U.S. Department of Health and Human Services announced that glass fiber would be listed as a material "reasonably anticipated" to be a carcinogen

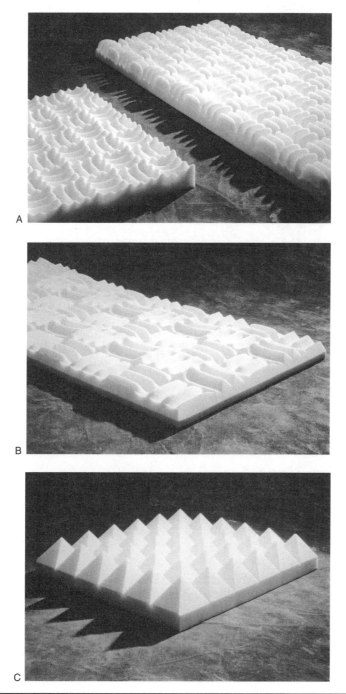

FIGURE 3-18   Three examples of foam panels manufactured by Sonex. (A) SONEXclassic. (B) SONEXone. (C) SONEXpyramids. (*Pinta Acoustic, Inc.*)

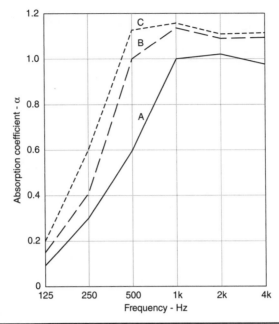

FIGURE 3-19   Absorption characteristics of SONEXclassic (polyester urethane foam, 2 lb/ft^3) with thicknesses of: (A) 2 in; (B) 3 in; (C) 4 in. The absorption coefficient theoretical maximum of 1.0 is exceeded because of the area effect. (*Pinta Acoustic, Inc.*)

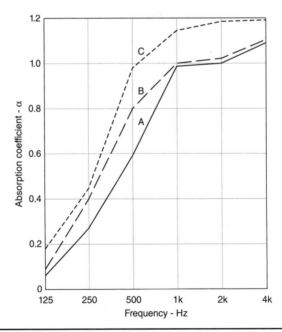

FIGURE 3-20   Absorption characteristics of SONEXpyramids (melamine foam, 0.7 lb/ft^3) with thicknesses of: (A) 2 in; (B) 3 in; (C) 4 in. The absorption coefficient theoretical maximum of 1.0 is exceeded because of the area effect. (*Pinta Acoustic, Inc.*)

(Hirschorn, 1994). Health issues relating to glass fiber have been controversial and hundreds of articles and reports have been published on the subject. Among the results were the following:

1. A 15-year epidemiological study found no excess of respiratory disease among glass fiber workers.

2. A two-year study at Los Alamos National Laboratory found no unusual disease patterns in laboratory animals after subjecting them to inhalation of glass fibers.

3. The World Health Organization concluded that glass fiber has not been a hazard to humans, but classified it as a "possible human carcinogen" because of studies that injected glass fibers into animals, which produced tumors.

Despite the lack of conclusive evidence incriminating glass fiber, the industry makes their products so that the glass fiber is contained. Owens-Corning has established an indemnification program to protect those involved in the "specification, application, and use" of their product. Glass fiber can cause skin, eye, and upper respiratory tract irritation. Proper precautions should be taken whenever handling glass-fiber products. This is also true of other man-made mineral fibers. It should be noted that glass fiber is distinctly different from asbestos fiber, which presents a serious health hazard.

CHAPTER 4

Diffusing Materials and Structures

Diffusion, in the casual acoustical sense, is the distribution of sound in a room. Diffusion is sometimes described as a scattering of sound. It can be compared to the effect that frosted glass has on light. More precisely, sound is perfectly diffuse when the energy density of the sound is uniform throughout a room (sound is homogeneous). Also, in a perfectly diffuse sound field, all directions of energy flow are equally probable (sound is isotropic). Generally, a high degree of diffusion is desirable in music performance rooms such as recording studios and concert halls; it enables sounds from musical instruments to be evenly distributed throughout the room. Diffusion is somewhat less important in sound playback rooms such as control rooms and home listening rooms where parameters such as imaging require greater localization and hence less diffusion.

In the past, reflection and absorption of sound were the most important variables in the acoustical design of an audio space. More recently, diffusion has assumed a greater role in acoustical design; in the room designs of this book, diffusion takes on its rightful prominence. When absorption is used, sound energy is decreased. When energy is reflected in a specular manner, the energy remains the same, but it is redirected in a new direction. When diffusion is used, sound energy remains the same, but it is scattered in many directions. Thus, diffusion offers a unique alternative to absorption and reflection.

There are many ways to cause diffusion of sound. In classical concert halls, whether or not the designers were consciously aware of it, the ornate decoration and statuary diffused sound throughout the hall. Concert halls with a cleaner, more modern design often lacked adequate diffusion and sometimes sounded poor because of it. The science of acoustical diffusion improved as the mathematics of diffusion became more clearly understood. Today, structures such as quadratic residue diffusers, described in this chapter, provide highly efficient diffusion and are a powerful tool in the acoustician's toolbox.

Sound Diffusion

Diffusion occurs when sound strikes an irregular surface and returns from it. However, unlike reflection from a flat surface that produces a "mirror" image of the sound, a diffused sound is scattered in many directions. This occurs because the different depths of the irregular surface create phase shifts in the returning sound. Diffusion may seem somewhat harder to conceptualize than reflection. But, both phenomena have much in common. As with reflection, diffusion is wavelength dependent. Sound at a certain frequency is only diffused from surface depths of a particular size. That is, for diffusion, the surface depths of a hard reflector must be similar to the wavelength of the sound. This is shown in the expression:

$$x > \lambda \qquad\qquad\qquad (4\text{-}1)$$

where x = surface depth, ft
  λ = wavelength of sound, ft

For example, a 1-kHz sine wave is about 1.1-ft long; thus it will be diffused from a hard surface with surface depths with dimension greater than 1.1 ft. Parenthetically, note that λ = speed of sound/frequency; the speed of sound is 1130 ft/sec. The diffused sound will travel throughout the room in many random directions. Diffusing panels contain a variety of depths that will diffuse sound over a broad frequency range. Importantly, as described below, certain geometries of surface depths can optimize diffusion.

Diffraction is another phenomenon that is related to reflection and diffusion. Diffraction is the bending of sound around an obstacle or through an opening. This explains why you can hear sounds from sources you cannot see. For example, sound bends around corners and travels down a corridor from one room to another. Diffraction is frequency dependent. Sound will diffract if the wavelength of the sound is larger than the obstacle. (Referring to the expression above, for diffraction, $x < \lambda$.) Thus lower frequencies are more easily diffracted than higher frequencies.

Low-Frequency Diffusion

Perfect diffusion calls for a sound field with uniform intensity everywhere in the room, and energy that flows equally in all directions. Perfect diffusion rarely occurs, and it is particularly difficult to achieve in small rooms. As we will see in Chap. 5, rooms have standing waves that distribute sound very unequally and also dictate that it flow in certain directions. These standing waves, also called room modes, occur at wavelengths that are comparable to the room's dimensions. Thus in small rooms, the standing waves occur at audible bass frequencies.

A lack of diffusion yields fluctuations in a room's reverberation response. When low-frequency reverberation is measured in small rooms, the decay characteristic is often irregular; the standing waves degrade the positive effects of any diffusion. This problem is difficult to overcome because there is no easy way to eliminate the effect of standing waves; at best, bass trapping can be used. Also contributing to the problem, as we have seen, is the fact that diffusion relates to the size of the diffusing surface. Very large surface depths are needed for low-frequency diffusion and these are difficult to supply in small rooms. In short, it is challenging to obtain adequate diffusion and hence smooth low-frequency reverberation in small rooms. At higher frequencies, the density of standing waves increases dramatically so sound becomes more homogeneous and isotropic. Also, at higher frequencies, smaller surfaces can be used to create diffusion.

Role of Diffusers in Room Design

Room surfaces can be treated by adding materials with different characteristics of absorption, reflection, and diffusion. Materials that either absorb or reflect in varying degrees can solve many acoustical problems; in either case, their use influences the liveness or deadness of a space. In some cases, absorption would solve the problem, but the resulting decrease in sound energy is undesirable. In that case, diffusion might be a better solution because it can overcome a problem due to reflection, and not decrease sound energy. For example, a flat surface reflects sound in a specular manner; the reflecting sound can combine with the direct sound (or another reflecting sound) resulting in constructive and destructive interference. This interference works like a comb filter to yield an uneven frequency response with peaks and dips. The effects of comb filtering are described in more detail in Chap. 15.

Sound reflects from a diffuser, but it is scattered in time and direction; when it combines with the direct sound, the regularity of a comb-filter response does not occur so the potential for audible distortion is reduced. Treatments can be devised that combine more than one characteristic. For example, a treatment might combine diffusion and absorption to prevent comb filtering, but also reduce sound energy. Also, as with any treatment, the operating frequency of a diffuser must be considered; as we have seen, the size of the diffuser determines the range of frequencies over which the diffuser is effective.

A room's acoustical characteristics can be examined with its impulse response; a short pulse (like a gunshot) is sounded, and the room's reverberant decay is plotted. Figure 4-1(A) shows the impulse response taken at the listening position in a small home theater before it is treated with absorbers and diffusers; a principal concern is sound quality from the front loudspeakers. As shown in the plot, before treatment is applied, there are distinct first-order early reflections (from the side walls and ceiling) as well as late reflections (from the rear wall). The early reflections are very harmful; they will combine with the direct sound at the listening position to yield a comb-filter frequency response. In addition, these early reflections will degrade front loudspeaker imaging in this listening room. The late reflections are isolated and yield an uneven ambient response as well as frequency-response anomalies.

Figure 4-1(B) shows the impulse response taken at the listening position after the room has been treated with absorbers and diffusers. After absorption is placed on the side walls and ceilings, the early reflections are eliminated. This will restore the loudspeakers' unaltered frequency response at the listening position and provide accurate imaging. Alternatively, instead of absorbers, diffusers could be applied to the front of the room. After diffusers are placed on the rear wall, the late reflections occur in a diffuse pattern that is similar to the reverberation found in a large room; this provides the listener with a sense of envelopment. Note the initial time-delay gap between the arrival of the direct sound and the late reflections. The gap is less than 20 msec; the late arriving sound will be psychoacoustically perceived as coming from the front of the room. This effect is very desirable.

This room design, with an absorptive front end and live (and diffusing) rear end, is sometimes called a live end-dead end (LEDE) room. One might ask exactly where the front absorbers should be placed. To resolve this, consider the walls and ceiling as a mirror; if you can see the sound sources (the front loudspeakers) in the "mirror" from the listening position, then that is a place where an early reflection can occur, and thus

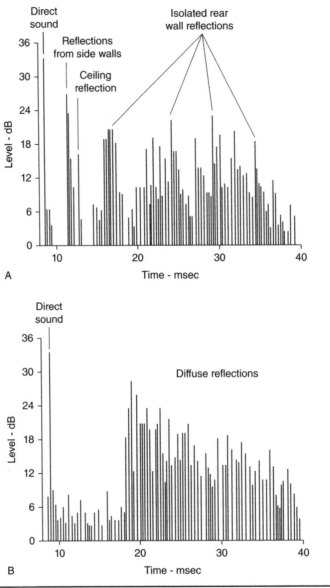

Figure 4-1 The impulse response of a small listening room. (A) Before treatment. (B) After treatment with diffusers.

should be covered with absorption. Instead of using absorbers, early reflections can be minimized or eliminated by angling the front walls and ceiling so that reflections are directed away from the listening position, and toward the rear of the room where sound strikes the diffusers. The listening position is said to be in a reflection-free zone (RFZ) because the effect of early reflections has been reduced. LEDE and RFZ are discussed in more detail in Chap. 15.

The LEDE design approach stresses the need to reduce specular reflections from the front loudspeakers. In a home theater with surround-sound playback, the problem is more complicated, but the same principles may be applied. Instead of segregating absorption and diffusion elements in the room, they could be mixed. For example, absorbers could be placed in the corners and diffusers on the middle wall surfaces and ceiling. Also, some hybrid absorber/diffuser modules are commercially available. Even then, one might choose to place more diffusion in the back of the room. This is because in most playback situations, the rear channels provide ambient sound, and this is expedited by diffusion on the rear surfaces.

To achieve the full benefits of a diffuser, the listener should not sit too close to it. In fact, the listener should be as far from the diffuser as possible. Generally, it is recommended that one should be at least three wavelengths away from the diffuser. For example, if the low-frequency limit of the diffuser is 500 Hz, the listener should be at least 7 ft away. In some room configurations, this may not be possible, so diffusion at the listening position is less than optimal. For example, some comb filtering may be heard in the frequency response.

When properly implemented, diffusers can solve many acoustical problems caused by specular reflections; moreover, they can do this without adding absorption. Because of the scattering effect, diffusers also make absorbers operate more efficiently. Diffusers can also give reverberant sound a sense of spaciousness and ambience.

Diffusion by Geometric Shapes

Depending on wavelength, sound may be reflected or diffused by protuberances of various shapes and sizes, such as those illustrated in Fig. 4-2. The flat wall surface of Fig. 4-2(A) reflects sound in a specular manner, with the angle of incidence equal to the angle of reflection. The triangular shape of Fig. 4-2(B) and the rectangular shapes of Fig. 4-2(C) reflect incident sound in various directions. This diffuses sound, but to a very limited degree. As noted above, to be effective, such irregularities must be large compared to the wavelength of the incident sound. Architects might find triangular, rectangular, or other shapes useful as sound diffusers and visual features in large concert halls, but space limitations and limited effectiveness tend to rule them out for use in smaller rooms.

Any sharp edge will diffuse sound to a certain extent, in addition to reflecting sound in various directions depending upon its shape. In this sense, there is modest true diffusion from edges such as found in door openings, wall corners, and even loudspeaker cabinets. The concave shape of Fig. 4-2(D) concentrates the sound at a focal point. Concentrating sound is the opposite of diffusing sound; hence, concave shapes should be avoided in most room designs.

The Polycylindrical Diffuser

The first radio and recording studios used only drapes, carpet, and acoustical tile to control the reverberation time of the space. This resulted in over-absorption of high-frequency energy, and under-absorption of low-frequency energy. The situation was favorably relieved through the increased popularity of polycylindrical diffusers, widely used in the 1940s and 1950s. At that time, some of the largest and best radio and recording studios incorporated a full complement of polycylindrical elements on walls and ceilings.

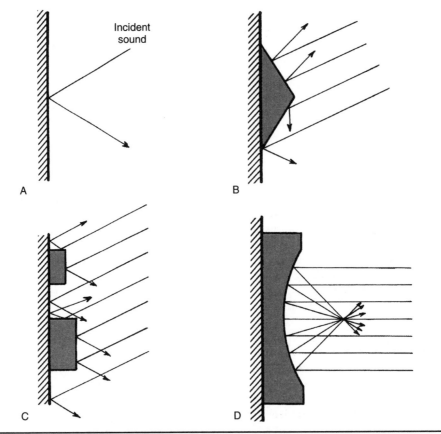

Figure 4-2 Patterns of reflection from different protrusion shapes.

Polycylindrical diffusers are not often seen in modern studios, although there is a trend toward their return. Their dramatic visual impact, low cost, and favorable acoustical characteristics are in their favor.

The polycylindrical surface shown in Fig. 4-3 comprises, for example, a curved wood panel with an airspace behind it. Three things can happen to a plane wave of sound striking such a cylindrical surface. In particular, some of the sound is absorbed, some sound is reflected, and some sound is reradiated.

That portion of the sound that is absorbed is largely in the low-frequency region, because of the usual large dimensions of the panel. Absorption at lower frequencies is much needed in some rooms, and can be difficult to achieve. The space between the curved diaphragm of a polycylindrical diffuser and the wall on which it is mounted is usually segmented into smaller spaces of varying volume. Each space is resonant at some low frequency, at which absorption peaks. As the sections are of different volumes, these absorption peaks will be spread over a range of low frequencies that together provide absorption in the often difficult low-frequency region.

That portion of the incident sound that is reflected from the polycylindrical face is dispersed in various directions. A polycylindrical face acts as a diffuser in one plane,

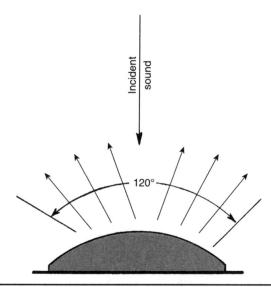

FIGURE 4-3   A polycylindrical diffuser showing reradiated sound.

and a specular reflector in another. For this reason, in some room designs, they are mounted both vertically and horizontally. This reflected component of dispersed sound must not be confused with the reradiated portion, although both disperse sound through a wide angle.

The reradiated portion of the sound results from the vibration of the curved diaphragm excited by the incident sound. This vibration of the polycylindrical diaphragm results in the reradiation back into the room of some of the sound energy. This reradiated sound does not follow the regular reflection law of equal angles of incidence and reflection. Rather, it is radiated almost equally throughout an angle of about 120° (Volkmann, 1941). A similar flat diaphragm radiates sound through a much smaller angle. The excessive use of sound-absorbing material to correct acoustical deficiencies of a space results in the loss of signal energy. With polycylindrical diffusers, both the reflected and the reradiated components help conserve the signal energy.

Polycylindrical diffusers are constructed by securing a thin sheet of plywood or Masonite over bulkheads shaped in semicircular segments. The sheets can be 3/16-in thick, and bulkheads are curved across a chord of 48 in, the standard width of plywood. To prevent rattles, a strip of foam, felt, or sealant is placed along the bulkhead before the skin is nailed into place. The airspace inside the skins is filled with glass fiber.

Reflection Phase Gratings

Diffraction gratings have long been the physicist's tool for studying light. The rainbow of colors produced by a beam of sunlight falling on a diffraction grating or prism is a familiar sight. Edwin Hubble, at Mt. Wilson Observatory, used a diffraction grating made of a strip of glass on which a great number of closely spaced lines had been engraved. By allowing the light from distant stars to fall upon this grating, he noted that the frequency of red light from a star varied with the distance to the star. From this observation, he originated the concept of the expanding universe.

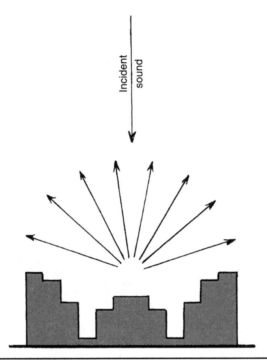

FIGURE 4-4 Diffusion of sound from a quadratic-residue diffuser.

Acoustical gratings provide a highly effective means of diffusing sound. The grating diffuser has eclipsed other means of diffusing sound in an enclosed space. Manfred Schroeder (Schroeder, 1975, 1979, 1988) is credited with the idea that grooved surfaces can make very effective sound diffusers. Acoustical gratings act as a diffuser in one plane, and a specular reflector in another; in some room designs, they are mounted both vertically and horizontally.

The theory of the reflection phase grating sound diffuser is based on number theory to obtain the critical sequence of groove or well depths that diffuse incident sound through wide angles. Schroeder first experimented with maximum-length sequences, but quadratic residue, primitive (read prime) root, and other sequences have been shown to have superior properties. In either case, a phase-grating diffuser uses a series of wells usually separated by thin dividers. Sound that strikes the phase-grating surface is diffused over a wide angle, as shown in Fig. 4-4.

Specular Reflection Theory

Sound striking a planar surface reflects from the surface as a specular reflection, that is, essentially in only one direction. Why is this? Consider that all points on a flat surface act like spherical emitters as shown in Fig. 4-5(A). The sound of all of these emitters is coherently combined to produce constructive interference (in which waves combine) and destructive interference (in which waves cancel). In the case of specular reflection, all scattered waves leaving the surface at an angle equal to the incident angle are

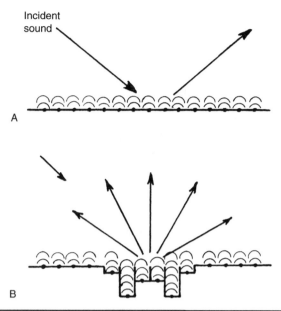

FIGURE 4-5 A graphical explanation of reflection and diffusion showing spherical emitters. (A) Specular reflection from a planar surface. (B) Diffusion from a phase grating.

relatively reinforced. This occurs because the specular waves differ in phase by integral numbers of wavelengths and hence they add. In other nonspecular directions, the waves cancel. Sound is actually scattered in all directions; the specular ones survive the interference process, and the nonspecular ones do not.

Reflection Phase Grating Theory

How do reflection phase-grating diffusers diffuse sound? As with specular reflection, the irradiated surface can be considered as covered with point-source spherical emitters of reflected energy. Unlike a planar reflector, a diffusing surface alters phases of the scattered sound so that it is constructively scattered in many directions, not just the specular direction. For example, a prime 7 quadratic-residue reflector (described below) can be sunk into the surface as shown in Fig. 4-5(B). The phases of the energy over the diffuser wells are different from that of the surrounding points because of the different depths of the wells. It takes time for sound to be reflected from the bottoms of the wells, and time is equivalent to phase. With the different phases at the different wells, the energy that was directed only to the specular direction in the case of a reflector is now directed into many directions uniformly. The diffraction directions are determined by the width of the repeat unit, not the depth sequence. All of the wells are of the same width.

To clarify the definition of terms, two periods of a typical diffuser are shown in Fig. 4-6. Separators are placed between wells for maximum efficiency, especially when operating at oblique angles, as they preserve the acoustical integrity of the wells. The incident sound enters the "top" of the well.

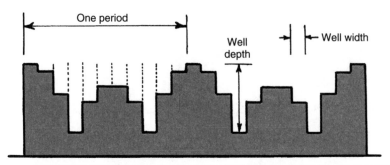

FIGURE 4-6 A graphical definition of terms used to specify a reflection phase-grating diffuser.

Reflection phase gratings operate over a limited frequency range. The low-frequency limit is imposed by the maximum depth of the wells, and the high-frequency limit is related to the width of the wells (D'Antonio and Konnert, 1984). The well width must be a half wavelength at the highest frequency to be effective as a diffuser, and the greatest well depth must be about 1-1/2 wavelength at the lowest frequency to provide adequate diffusion. For example, a maximum well depth of 12 in and width of 1 in yields a frequency range from 323 Hz to 5780 Hz. Experience has shown that the effective range is roughly about half an octave lower and half an octave higher than such figures.

Reflection phase-grating diffusers are also absorbers. They provide useful low-frequency absorption below the lower limit of diffusion. With wells of varied depth, sound absorption occurs because of increased particle velocity flows from one well to another to equalize sound pressure on the face of the diffuser. This mechanism provides serendipitous low-frequency absorption.

An overview of the mechanisms of sound absorption, reflection, and diffusion is shown in Fig. 4-7. The three kinds of acoustical treatments are shown [Fig. 4-7(A)]. Much of the sound falling on an absorbing material is absorbed, but a small amount is reflected. In the temporal response column [Fig. 4-7(B)], the reflected sound is attenuated about 20 dB. In the spatial response column [Fig. 4-7(C)], the angle of the reflection is equal to the angle of the incident ray, but the magnitude of the reflection is reduced by absorption.

The reflection from a hard surface has almost the same amplitude as the incident sound. The reflection from the reflection phase-grating diffuser, however, has several important characteristics:

1. The diffused energy is spread over time, as shown by the temporal response.

2. The amplitude of the diffused return is attenuated approximately 8 to 10 dB.

3. The sound is scattered through the entire half circle, as shown by the spatial response.

These three points have great practical value, and will be referred to many times.

Figure 4-8(A) shows the hemidisc of sound diffused from the array of wells of a quadratic-residue diffuser. Figure 4-8(B) reminds us that sound arriving from another plane is reflected from the face of a phase-grating diffuser in a specular manner, that is, as though it had a flat facing. Note that the plane of specular reflection follows the rule that the angle of incidence equals the angle of reflection. One of the advantages of the

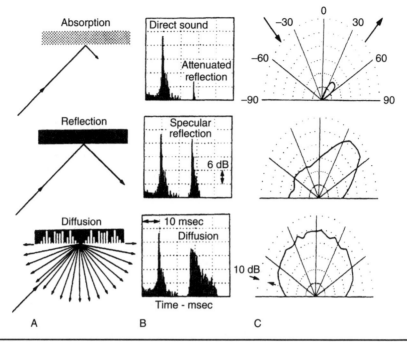

FIGURE 4-7   A comparison of sound incident on absorptive, reflective, and diffusive acoustical surfaces. (A) Acoustical treatment. (B) Temporal response. (C) Spatial response.

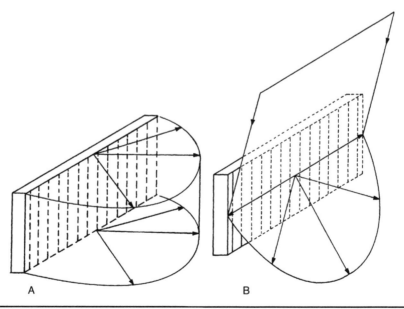

FIGURE 4-8   Horizontal and vertical reflections from a hemidisc diffuser face. (A) Diffusive scattering. (B) Specular reflection.

primitive-root diffuser is that the specular scattering at the design frequency and its multiples are suppressed. This specular component exists and sometimes might cause problems, but in most installations it simply adds to the diffusion of sound in the room.

The Quadratic-Residue Diffuser

Diffuser designs can be based on several different number theory sequences that define the well geometry. Of the various approaches, the quadratic-residue sequence with its natural symmetry has proved to be of the greatest practical value. The maximum well depth determines the longest wavelength of sound to be diffused. The well width is about a half wavelength at the shortest wavelength to be scattered. The relative depths of a sequence of wells are found from the equation:

$$\text{Well depth proportionality factor} = n^2 \text{ modulo } p \tag{4-2}$$

where p = a prime number
$\quad\quad n$ = an integer ≥ 0

A prime number is any number that is not divisible without a remainder by any other integer (exceptions: +1, −1, and the integer itself). Examples of primes are 5, 7, 11, 13, 19, 23, etc. The modulo refers to residue. For example, inserting $p = 11$ and $n = 7$ into Eq. 4-2 gives 49 modulo 11. Modulo 11 means that 11 is subtracted successively from 49 until the significant residue 5 is obtained. That is simple, but higher values such as $n = 15$ become awkward. To simplify calculations, many calculators and programs have a modulo function. Figure 4-9 shows quadratic-residue profiles for p from 7, 11, 13, and 17.

The Primitive-Root Diffuser

Diffusers can also be constructed according to a primitive-root sequence. The number theory sequence for primitive-root diffusers is:

$$\text{Well depth proportionality factor} = g^n \text{ modulo } p \tag{4-3}$$

where p = a prime number
$\quad\quad g$ = the least primitive root of p
$\quad\quad n$ = an integer ≥ 0

Figure 4-10 shows primitive-root profiles for three combinations of p and g. Primitive-root diffusers lack symmetry, which can be an advantage in some circumstances and a disadvantage in others. Primitive-root diffusers, as with other welled surfaces, reduce the energy returned by a specular reflection at the design frequency and its multiples.

Fractals

The self-similarity property of fractals has been applied to quadratic-residue diffusers to extend their frequency range and minimize lobing for large-area coverage. There are certain manufacturing constraints that limit how narrow or how deep a diffuser

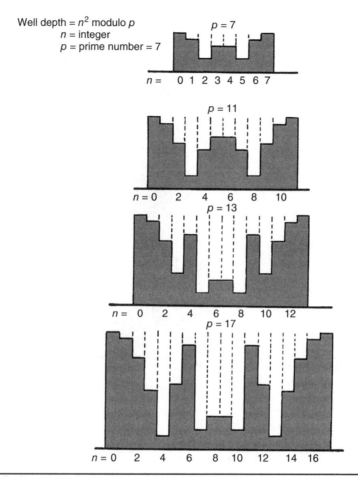

Well depth = n^2 modulo p
n = integer
p = prime number = 7

FIGURE 4-9   Profiles of quadratic-residue diffusers.

well can be made. A similar problem with loudspeakers has been solved by woofers to extend the low-frequency radiation, and tweeters to extend the high-frequency radiation. Normal quadratic-residue diffusers can be designed for any reasonable low-frequency limit.

The fractal principle of self-similarity is used to extend the high-frequency limit. The general process is illustrated in Fig. 4-11. The upper left quadratic-residue unit is built on the prime 7. To extend its high-frequency range, another small prime-7 diffuser (circle #2) is fitted into the bottom of the indicated well (circle #1). In addition, smaller diffusers of the #2 type are fitted into every other well bottom of #1. Thus diffuser #1 acts in a normal way, except that the smaller #2 units in each well result in much greater high-frequency diffraction. Moreover, an even smaller prime-7 diffuser (circle #3) can be fitted into the bottom of each well of diffuser #2. This yields diffusers within diffusers within diffusers, each covering a different frequency range. A diffuser 16 ft long, 6 ft 8 in high, and 3 ft deep offers a bandwidth from 100 Hz to 17 kHz using the fractal principle.

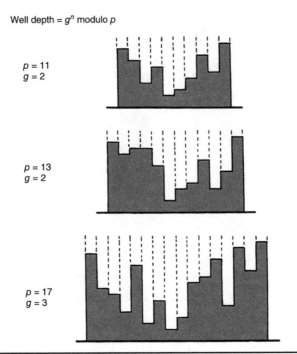

Well depth = g^n modulo p

$p = 11$
$g = 2$

$p = 13$
$g = 2$

$p = 17$
$g = 3$

FIGURE **4-10** Profiles of primitive-root diffusers.

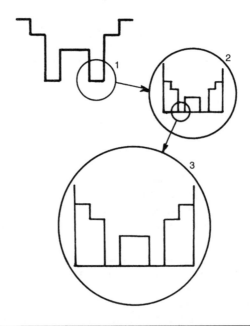

FIGURE **4-11** The fractal principle can be applied to diffusing surfaces.

The Diffusion Coefficient

Diffuser performance can be quantified with the normalized diffusion coefficient. It measures how uniformly a diffuser distributes sound over a semicircular arc compared to a flat surface of the same size. It is a normalized value because it corrects for the effect of the reflector that it is being compared to. In this way, the coefficient measures only the diffuser. The diffusion coefficient is the ratio of the scattered intensity at ±45° to the specular intensity. The coefficient ranges from 0 to 1.0; higher values indicate better diffusion.

Commercially Available Reflection Phase-Grating Diffusers

As noted, credit for the original idea of the reflection phase-grating acoustical diffuser goes to Manfred Schroeder. His insight into the mathematics upon which such diffusers are based and his knack for sensing practical uses are profound. Credit for commercialization of these ideas goes to Peter D'Antonio, a physicist with scientific research experience. He founded RPG Diffusor Systems, Inc. and began supplying practical and efficient diffusing elements to audio, recording, and architectural industries. The studio designs of this book owe much to the availability of commercial reflection phase-grating diffusers and to the generous cooperation of Dr. D'Antonio.

The RPG products described below are commercially available; they are used here because of the comprehensive specifications and laboratory measurements that are available with them. However, the physics underlying their designs is freely available to anyone. Diffusers adhering to Schroeder's theories can be designed and constructed by anyone with the necessary knowledge and carpentry skills. Depending on many factors, this may or may not be more cost efficient than purchasing commercially available units.

The QRD-734 Quadratic-Residue Diffuser

The QRD-734 line of quadratic-residue diffusers (manufactured by RPG Diffusor Systems, Inc.) is shown in Fig. 4-12. These diffusers are based on the prime 7, and are available in nominal dimensions of 4 × 4 ft, 2 × 4 ft, 4 × 2 ft, and 2 × 2 ft with a depth of 9 in. The width of the wells is 2-1/2 in. The well dividers are 1/2 in. As with other phase-grating diffusers, when the module is placed with its wells aligned

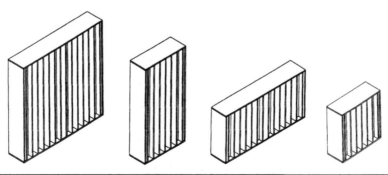

FIGURE 4-12   Examples of QRD-734 diffuser modules.

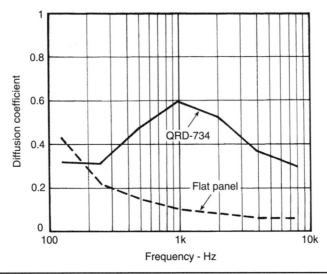

FIGURE 4-13 The diffusion coefficient of QRD-734 diffuser modules.

horizontally, diffusion is vertical; conversely, vertical wells generate diffusion horizontally. The QRD-734 also offers sound absorption coefficients varying in the range of 0.2 to 0.3.

The performance of QRD-734 diffusers is shown in Fig. 4-13. The graph shows the diffusion coefficient versus frequency; this plot suggests excellent diffusion characteristics. In this graph, the diffuser is compared to a flat panel in regard to diffusion effectiveness. This graph shows the true bandwidth of this particular diffuser. It is generally easier to diffuse high-frequency sound than low-frequency sound, but the modules have been designed to cover a wide range of the audible band.

The Formedffusor

The Formedffusor, shown in Fig. 4-14, is a lightweight unit capable of providing broadband, uniform sound diffusion in either the horizontal or vertical planes for all angles of incidence by proper orientation of the units. The Formedffusor is used in the standard T system as a suspended ceiling. It gives good mid- to low-frequency diaphragmatic sound absorption as shown in Fig. 4-15. The diffuser is formed from Kydex by a thermoforming process so uniform wall thickness is assured.

The FRG Omniffusor

The FRG Omniffusor, shown in Fig. 4-16, provides broadband, uniform sound diffusion simultaneously in both horizontal and vertical planes for all angles of incidence. It is based on the two-dimensional quadratic-residue sequence with prime 7. The FRG Omniffusor is made of fiberglass-reinforced gypsum (FRG). This two-dimensional diffuser provides uniform hemispherical coverage from 500 to 3000 Hz. The nominal size of each unit is 2 ft × 2 ft × 5 in.

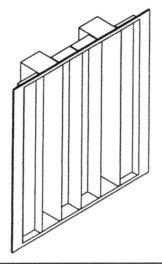

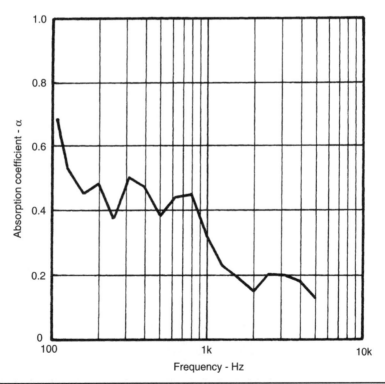

FIGURE **4-15**   Low-frequency absorption of the Formedffusor diffuser module.

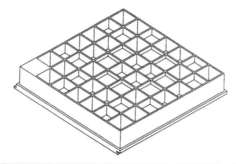

FIGURE 4-16   The Omniffusor diffuser module.

The FlutterFree Diffuser

The FlutterFree diffuser, shown in Fig. 4-17(A) and (B), is a long plank of hardwood routed to form a 7 prime quadratic-residue diffuser. Available in 4-ft or 8-ft lengths, the width of each panel is 4 in and the thickness is 1 in.

As we observed in Chap. 2, our hearing is very sensitive to flutter echoes produced by successive reflections from opposing parallel surfaces. Patches of absorbent material

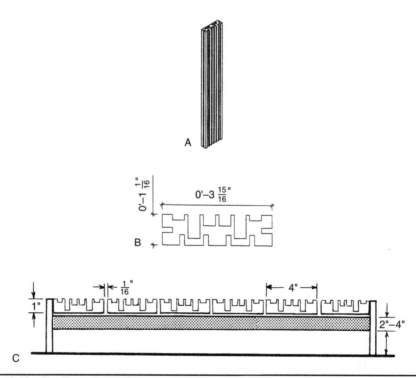

FIGURE 4-17   The FlutterFree diffuser module. (A) Isometric view. (B) Plan view. (C) A Helmholtz absorber with FlutterFree slats.

can be used to eliminate this echo. However, in some cases, the loss of sound energy from added absorption is undesirable. As an alternative, diffusing panels such as FlutterFree strips can be used to eliminate flutter echoes without adding absorption to the room.

Sound absorbers of the Helmholtz resonator type can utilize FlutterFree slats. Such a low-frequency sound absorber is shown in Fig. 4-17(C). The strips of FlutterFree can be butt-joined with Lamello splines, or joined together with a space of approximately 1-1/8 in with Lamello spacers.

CHAPTER 5

Room Modes and Room Geometry

One of the most important aspects of a room's acoustical characteristics is the way that sound waves interact with each other as they reflect from the interior surfaces of the room. Energy radiated from a sound source will travel throughout a room and be reflected from the interior surfaces. For example, in a rectangular room, a sound wave might reflect between two opposing surfaces (such as two walls or the floor/ceiling), between four surfaces, or between all six surfaces. These reflection patterns establish room modes and in particular three kinds of modes known as axial, tangential, and oblique respectively.

As sound reflects back and forth between the surfaces, each mode exists as a standing wave that is static between the room surfaces with a pressure maxima, pressure minima, and varying pressure in between. A listener can walk across the room and hear the sound level change along different points on the standing wave. Each mode exists at a fundamental frequency and at frequencies that are multiples of the fundamental. The fundamental frequency of each mode is determined by the physical dimensions of the room and specifically the distances between each opposing surface.

These modes are easy to calculate in a six-sided rectangular room with parallel surfaces. A room with a different geometry, for example, with splayed walls, also supports modes but the physical placement of the pressure response of the modes is distorted by the nonparallel surfaces. In other words, splaying walls does not solve mode problems; it only makes them harder to predict.

Room modes present a challenge to acoustical design because they distribute sound energy very unequally in a room. At a given location, certain frequencies may be boosted or cut, yielding an irregular frequency response. At another location, the response may be quite different. This affects sound radiating from a musical instrument and sound received at a listening position. The problem is particularly evident in small rooms because the short dimensions place the frequency-response irregularities in the audible low-frequency range.

Room Cutoff Frequency

As we noted in Chap. 1, small rooms pose unique problems that do not affect large rooms. Ideally, we could trace the path of each ray of sound and, with the help of sound images on the various surfaces, compute the sound pressure at the ear of the listener. That ideal condition is approached in a large room, and in a small room above a certain cutoff (or crossover) frequency. The cutoff frequency helps establish the transition zone in a room between large- and small-room analyses. The cutoff frequency can be estimated from the room's reverberation time and volume:

$$f_c = 11,885 \sqrt{\frac{RT_{60}}{V}} \qquad (5\text{-}1)$$

where f_c = room cutoff frequency, Hz
RT_{60} = reverberation time, sec
V = volume of the room, ft^3

When the volume of a room is large compared to the wavelength of the sound in it, that is, for frequencies higher than the cutoff frequency, the concept of sound rays has validity. For example, at 5000 Hz, the wavelength of sound is 1130/5000 = 0.23 ft. At this frequency, the length of a room could be of the order of 100 wavelengths, and the ray concept applies. However, a wave approach is required for studying the modal resonances of small rooms, that is, rooms the size of many recording studios and listening rooms. For example, at 20 Hz, the wavelength is 1130/20 = 56.5 ft, which is far less than the cutoff frequency and greater than the length of the average studio.

For example, in a studio of 3000 ft^3 with a reverberation time of 0.5 sec, the cutoff frequency would be about 150 Hz. For a studio twice the volume (but the same RT_{60}), the cutoff frequency would be about 110 Hz, and for a studio half the volume it would be about 215 Hz. Below the cutoff frequency, the ray concept of sound has little meaning; instead, room-mode analysis must be used. (Conversely, above the cutoff frequency, because modes are so densely spaced, room-mode analysis cannot be used.)

In rooms the size of most recording studios, listening rooms, and control rooms, the low-frequency acoustics of the space is dominated by room-mode effects. These room-mode resonances control the low-frequency response of the room and as a result, the response is often very uneven; response is boosted where modes are reinforced, and cut where modes are canceled. An example of the low-frequency response of a small room is shown in Fig. 5-1.

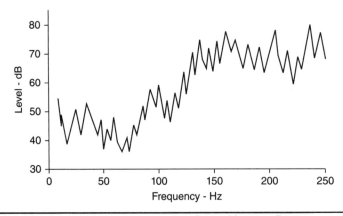

Figure 5-1 Example of the low-frequency response of sound in a small room.

Music excites these low-frequency room resonances in a most transient and variable way as a result of the time fluctuations of the music signal. These resonances of the room must be studied because they dictate the sound field of the room for the lower part of the audible spectrum. At higher frequencies, the modes are more closely spaced and their individual effect at specific frequencies is diminished; as a result, the frequency response at higher frequencies is flatter and more consistent. Bass traps and panel absorbers can be used to control uneven low-frequency response.

Resonances in Tubes

Air-filled tubes or pipes (such as organ pipes) demonstrate how sound acts in confined spaces; they provide a simple model to help understand room modes. In each case, the dimensions of the container (tube or room) give rise to a set of certain preferred frequencies. The elementary association of tube length with the pitch of the resultant tone is intuitive. Tubes can have ends that are both closed, closed at one end, or open at both ends. No matter whether a tube has a closed end or not, that end is a reflector. It is easy to visualize a closed end as a reflector, but an open end is a reflector, too. The sound coming down the tube encounters a great change in impedance where the tube ends and the exterior space begins. That change in impedance reflects energy back toward the source. So, open or closed, the ends act as reflectors. However, the effect of open or closed ends provides somewhat different resonant conditions.

Because it is directly analogous to the surfaces of a room, let's consider the case of a tube closed at both ends, as shown in Fig. 5-2. A tube with a reflector on each end is an acoustical cavity in which standing waves can be set up and sustained. For example, sound traveling toward the right end of a tube is reflected from the end, and the reflected energy travels back in the opposite direction. At the same time, a reflection from the left

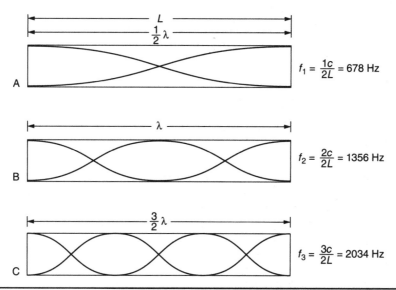

$$f_1 = \frac{1c}{2L} = 678 \text{ Hz}$$

$$f_2 = \frac{2c}{2L} = 1356 \text{ Hz}$$

$$f_3 = \frac{3c}{2L} = 2034 \text{ Hz}$$

Figure 5-2 Resonances in a tube closed at both ends. Sound-pressure level is shown.

end travels toward the right. The left-going and the right-going waves interact with each other in such a way that at certain frequencies a standing wave is set up in resonance.

We observe that the fundamental standing wave has a wavelength that is twice the length of the tube. (Recall that the relationship between wavelength and frequency is $\lambda = c/f$.) If the frequency of the sound exciting the tube is such that the length L of the tube is one-half wavelength long, a standing wave is set up and the pressure pattern within the tube would be as shown in Fig. 5-2 (A), with a node at the center and antinodes at each end. The node is at minimum sound pressure, and the antinodes are at maximum sound pressure. Alternatively, air-particle displacement could be plotted. Furthermore additional standing waves occur at multiples of this frequency. In particular, the resonant frequencies of a tube closed at both ends can be determined:

$$f_n = nc/2L \qquad (5\text{-}2)$$

where f_n = resonant frequencies, Hz
  n = an integer ≥ 1
  c = speed of sound = 1130 ft/sec
  L = length of tube, ft

For example, consider a tube closed at both ends and length of 10 in (0.83 ft). Its fundamental resonant frequency can be calculated:

$$f_1 = \frac{1c}{2L} = 678\,\text{Hz}$$

Another standing wave is set up at twice the fundamental frequency:

$$f_2 = \frac{2c}{2L} = 1356\,\text{Hz}$$

Another standing wave is set up at three times the fundamental frequency:

$$f_3 = \frac{3c}{2L} = 2034\,\text{Hz}$$

These three standing waves are shown in Fig. 5-2. In fact, a standing wave exists at every integral multiple of the fundamental resonant frequency. In a tube that is closed at both ends, no matter what resonant frequency, the sound pressure at each end of the tube is at maximum value.

There are significant similarities between the tube closed at both ends (Fig. 5-2) and the tube open at both ends as shown in Fig. 5-3. In both cases, a standing wave can exist for every integral multiple of the lowest resonant frequency. Also in both cases, the lowest resonant frequency corresponds to the tube's one-half wavelength length. However, in an open-ended tube, a node must exist at each end for the tube. In comparing Figs. 5-2 and 5-3, quite different pressure patterns within the tubes are noted, but the similarities outweigh the differences. The more basic fact is that for both ends open and both ends closed, standing waves can exist for integral multiples of the lowest resonant frequency, and are at identical frequencies.

The case of a tube with one end open and the other end closed, shown in Fig. 5-4, is quite different from both ends open or closed. Here resonances occur only at odd

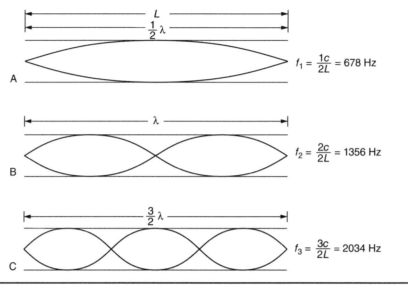

FIGURE 5-3 Resonances in a tube open at both ends. Sound-pressure level is shown.

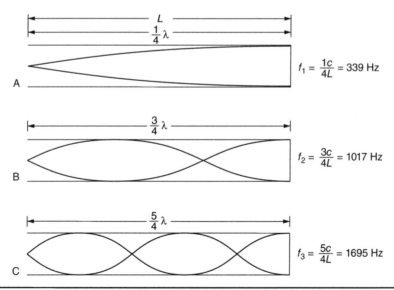

FIGURE 5-4 Resonances in a tube with one end closed, and one end open. Sound-pressure level is shown.

integral multiples of the lowest frequency, and at the lowest frequency the tube is one-quarter wavelength long.

Figures 5-2 through 5-4 plot sound-pressure level inside a tube. Alternatively, we could plot air-particle displacement. At any point, the particle displacement is exactly out of phase with the sound-pressure level; there is a displacement node at every pressure antinode, and a displacement antinode at every pressure node. For example, in a closed tube, particle displacement is always zero at both ends. For the lowest mode, air-particle displacement is maximum at the center point; this center position from the end of the tube corresponds to $\lambda/4$ for this mode. Similarly, for any mode, although the physical distance differs, the first position of maximum air-particle displacement corresponds to $\lambda/4$. This explains why in a closed tube, and in a room, porous absorbers are most effective when placed at a distance of $\lambda/4$ from a boundary.

In these examples the tube is equivalent to an enclosed space, irrespective of the end condition. These enclosed spaces resonate at some basic frequency and multiples of that basic frequency. Tubes and studios differ in many ways, but they are similar in that they are both resonant systems with standing waves forming at multiple frequencies.

A word about standing wave terminology is in order here. The physicist is apt to use the standard term "stationary wave" if losses are assumed to be zero. If there are terminal losses present, the ideal situation is disturbed and the term "standing wave" applies. In all our practical situations there will be losses, and the term "standing wave" is applicable.

Turning to the acoustics of rooms, the observer is inside the tube, so to speak. The two reflecting ends of the tube now become six surfaces, each of which reflects sound. As far as the simplest (axial) modes are concerned, the situation in the room is as though there is a horizontal tube, a vertical tube, and a crosswise tube. Clearly, fundamental similarities exist between the tube and the room. In particular, they behave the same, but whereas a tube operates across one dimension, a room operates across three dimensions.

Resonances in Rooms

A room can be considered as a resonating chamber. Consider first a pair of opposing, parallel wall surfaces as shown in Fig. 5-5. These two isolated walls act as a resonating system. As in a tube, a tonal sound emitted between the two walls travels to the left wall and is reflected toward the right wall, then is reflected back toward the left wall, and so on. The initial sound simultaneously travels also toward the right wall, is reflected to the left, and then to the right. The right-going and the left-going waves combine, and if the wavelength of the sound is properly related to the distance L between the two surfaces, a standing wave will be produced while the sound continues.

The first mode of this standing wave, at the fundamental frequency, has a maximum sound pressure at the surface of the walls, and a null midway between them. At twice that frequency, a second standing wave mode results with two nulls between the walls. At three times the frequency, in the third mode, three nulls appear. The sound pressure is maximum at the surfaces for all modes. These are called axial modes because the sound travels parallel to the axis normal to the two surfaces. Likewise, axial modes occur along each axis of the room, that is, the length, width, and height. As described below, there are also two other types of room modes, one based on tangential reflections, and the other on oblique reflections.

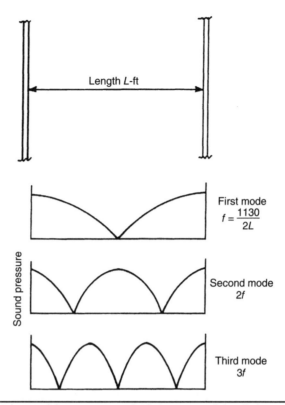

$$f = \frac{1130}{2L}$$

First mode

Second mode
$2f$

Third mode
$3f$

FIGURE 5-5 As in a tube, modal resonances exist between parallel walls in a room.

The sound level at the resonant frequency changes as a listener moves along a standing wave. This can be demonstrated with simple experiments as described below. Some frequencies will exhibit a low-frequency null in the center of the room. This can be problematic because the center of the room is often the place where a mixing engineer might sit behind a console or where a listener may sit in a listening room. This null is sometimes called "bass suckout" and should be avoided if possible. For example, opposite ends of a room should not both be highly reflective. Instead, for example, one wall should be absorptive.

In a room with an extremely absorptive wall or that is open on one side, the open dimension can be modeled as a tube that is closed at one end and open at the other; sound pressure is maximum at the closed end and minimum at the open end. The fundamental frequency is half of that in a closed-tube room and there is no minimum in the center of the room. Splaying opposing walls is useful for eliminating flutter echoes, but is not a solution for room mode problems. Room modes will still exist; the splaying only changes the symmetry of the modal patterns in the room. Choosing proper room proportions is a better (but incomplete) solution, which is discussed below. Alternatively, installation of Helmholtz resonators can decrease energy at problematic modal frequencies.

It is clear that the longer the tube, the lower the lowest resonant frequency will be. Likewise, larger rooms, with longer distances between opposing surfaces, will have a lower fundamental resonant frequency. In very large rooms such as concert halls, the lower resonant frequencies will be so low that they are not a concern. However, in small rooms (perhaps smaller than 1500 ft³), the lower resonant frequencies are relatively higher in frequency, and can thus affect higher, more critical frequency areas.

Axial, Tangential, and Oblique Modes

In a room, waves can travel backward and forward between any two opposing walls, that is, in an axial direction. Waves can also strike room surfaces at angles. At certain angles the waves return upon themselves and set up tangential and oblique standing waves. These are normal modes of vibration of the room, comparable to the normal modes of vibration in a tube.

Axial modes exist in each of a rectangular room's three dimensions, as shown in Fig. 5-6. All six surfaces of the room play a role, but each axial mode reflects between opposite and parallel walls. There is one set of axial modes created by reflections from the near and far end walls of the room. A second set of axial modes exists between the left and right side walls of the room. A third set of axial modes exists between the floor and the ceiling. Each set of axial modes has a lowest frequency based on twice the distance between the wall pairs (2L), as well as a series of integral multiple frequencies of 2L. When the distance between parallel surfaces is the same for more than one pair, or a multiple of another pair, axial modes (either fundamental or multiples) can appear at the same frequency. This accumulation of energy can create a peak in the room's frequency response. Conversely, dips in the room's frequency response can appear. To minimize the effect of these peaks and dips, a room's dimensions should be chosen so that mode frequencies are as evenly distributed as possible.

In addition to axial modes, rooms can also resonate in tangential and oblique modes, as shown in Fig. 5-7. Tangential modes occur for a wave that strikes four walls and comes back to its starting place. As with axial modes, tangential modes exist in each of the room's three dimensions and appear as a series of integral multiple frequencies (see Fig. 5-7). Oblique modes strike all six surfaces of a room each round trip. As with axial modes, oblique modes use all three room dimensions and have a series of integral multiple frequencies (see Fig. 5-7).

Axial, tangential, and oblique modes have different energy levels. Because they incur fewer boundary reflections, axial modes contain the most energy. Because axial modes are the more potent, an appraisal of axial modes alone might give a good estimate of the performance of a space under consideration. Because tangential modes strike four surfaces, they have less energy than axial modes, and are less consequential.

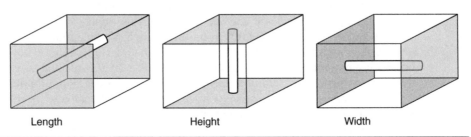

| Length | Height | Width |

FIGURE 5-6 Axial modes exist in each of a room's three dimensions.

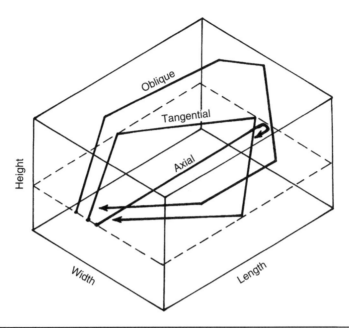

Figure 5-7   A comparison of axial, tangential, and oblique modes.

Likewise, oblique modes strike all six room surfaces, so they have less energy than either axial or tangential modes. (However, it should be noted that absorption is lower for such smaller angles of incidence compared to the perpendicular angle of incidence that occurs with axial modes.) In particular, tangential modes have only half the energy of the axial modes, which means they are reduced by 3 dB. Oblique modes, having only one-fourth the energy of axial modes, are reduced by 6 dB (Morse and Bolt, 1944).

Room Mode Equation

Room modes can be calculated with a formula that is similar to that used to calculate modes in a tube. For a rectangular room, there are three pairs of opposing, parallel surfaces. They are associated with the length L, width W, and height H of the room. The equation for calculating the frequencies of all the normal modes of resonance is:

$$f = \frac{c}{2}\sqrt{\frac{p^2}{L^2} + \frac{q^2}{W^2} + \frac{r^2}{H^2}} \tag{5-3}$$

where f = mode frequency, Hz
c = speed of sound = 1130 ft/sec
L, W, H = room length, width, height, ft
p, q, r = integers 0, 1, 2, 3, etc.

In an existing room, the room's length L, width W, and height H are fixed. In a new room design, L, W, and H can be varied; moreover, the values should be selected to yield more optimal modal results as described below. The integers p, q, and r are the only variable integers in the equation. As they are varied, they determine the frequency of each mode. They also provide a system of identification for every mode. For example, if $p = 1$, $q = 0$, and $r = 0$, the mode is identified as (1, 0, 0). Moreover, $p = 1$ refers to the first mode, $p = 2$ refers to the second mode, $p = 3$ refers to the third mode, and so on.

Table 5-1 lists and identifies all the modal frequencies for a small room measuring 12 ft, 5-1/2 in × 11 ft, 5 in × 7 ft, 11 in, extending to p, q, $r = 4$. Note that a frequency of 45.3 Hz is the lowest frequency supported by room resonance. Figure 5-8 plots the axial,

Mode Number	Integers p q r	Mode Frequency, Hz	Axial	Tangential	Oblique
1	1 0 0	45.3	x		
2	0 1 0	49.5	x		
3	1 1 0	67.1		x	
4	0 0 1	71.5	x		
5	1 0 1	84.7		x	
6	0 1 1	87.0		x	
7	2 0 0	90.7	x		
8	2 0 1	90.7		x	
9	1 1 1	98.1			x
10	0 2 0	98.9	x		
11	2 1 0	103.3		x	
12	1 2 0	108.8		x	
13	0 2 1	122.1		x	
14	0 1 2	122.1		x	
15	2 1 1	125.6			x
16	1 2 1	130.2			x
17	2 2 0	134.2		x	
18	3 0 0	136.0	x		
19	0 0 2	143.0	x		
20	3 1 0	144.8		x	
21	0 3 0	148.4	x		
22	2 2 1	152.1			x
23	3 0 1	153.7		x	
24	1 1 2	158.0			x
25	3 1 1	161.5			x
26	0 3 1	164.8		x	
27	3 2 0	168.2		x	
28	2 0 2	169.4		x	

TABLE 5-1 Mode Calculations for a Small Rectangular Room (Room Dimensions: 12 ft, 5.5 in × 11 ft, 5 in × 7 ft, 11 in)

Mode Number	Integers p q r	Mode Frequency, Hz	Axial	Tangential	Oblique
29	1 3 1	170.9			x
30	0 2 2	173.9		x	
31	2 3 0	173.9		x	
32	2 1 2	176.4			x
33	1 2 2	179.7			x
34	4 0 0	181.4	x		
35	3 2 1	182.8			x
36	2 3 1	188.1			x
37	2 2 2	196.2			x
38	0 4 0	197.9	x		
39	3 0 2	197.9		x	
40	3 3 0	201.3		x	
41	3 1 2	203.5			x
42	0 3 2	206.1		x	
43	1 3 2	211.1			x
44	0 0 3	214.6	x		
45	1 0 3	219.3		x	
46	0 1 3	220.2		x	
47	3 2 2	220.8			x
48	1 1 3	224.8			x
49	2 3 2	225.2			x
50	2 0 3	232.9		x	
51	4 3 0	234.4		x	
52	0 2 3	236.3		x	
53	2 1 3	238.1			x
54	3 4 0	240.2		x	
55	1 2 3	240.6			x
56	3 3 2	247.0			x
57	2 2 3	253.1			x
58	3 0 3	254.0		x	
59	0 3 3	260.9		x	
60	3 2 3	272.6			x
61	2 3 3	276.2			x
62	4 0 3	281.0		x	
63	0 0 4	286.1	x		
64	0 4 3	291.1		x	
65	3 0 4	316.8		x	
66	0 3 4	322.3		x	

TABLE 5-1   (Continued)

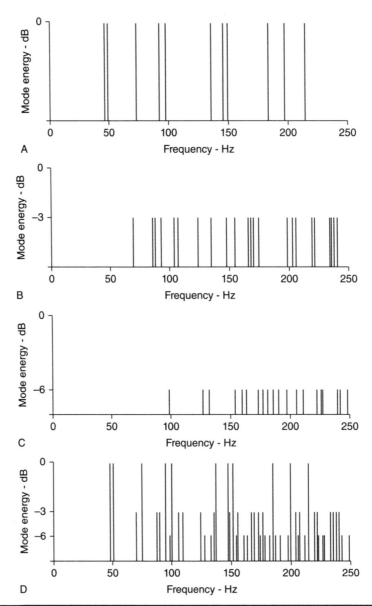

Figure 5-8   Plot of modes for the small room specified in Table 5-1. (A) Axial. (B) Tangential. (C) Oblique. (D) Combined.

tangential, and oblique modes of this room falling below 250 Hz. The height of the lines reflects the relative intensity of the modes (0, −3, −6 dB).

Figure 5-9 separately shows three modes in this room. In this figure, the p variable is set at 1, 2, and 3, yielding the first three axial modes along the length of the room.

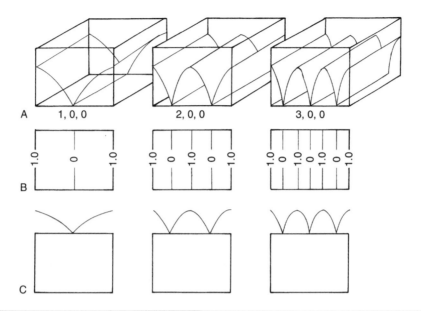

Figure 5-9 Plot of the first three axial modes along the length of a room. (A) Vectors of sound pressure rise from the floor to a height depending on the sound pressure at that point. (B) A flat-plane contour plot of sound pressure. (C) Plot of sound pressure on and above the floor.

The wavefront lines indicate planes of constant sound pressure extending from floor to ceiling. If the sound-pressure pattern of all the modal frequencies of Table 5-1 were depicted simultaneously, the plot would be extremely intricate. Combining only the (1, 0, 0) mode and the (2, 0, 0) mode would be daunting, let alone all those listed in the table. The physical sound field is complex, even in the simplest situation. As a loudspeaker plays the sound of music in a room, the sound pressure at any given point at any instant is the sum total of all these rapidly changing room modes.

Table 5-2 lists the axial, tangential, and oblique modes for a large room having the dimensions of 23 ft, 4 in × 16 ft × 10 ft. Figure 5-10 plots the axial, tangential, and oblique modes of this room falling below 250 Hz. It is interesting to compare this room's modes with those of the smaller (12 ft, 5-1/2 in × 11 ft, 5 in × 7 ft, 11 in) room shown in Fig. 5-8. The larger room has 65 modal frequencies below 250 Hz, while the smaller room only has 56 modes. In the larger room, the modes begin at lower frequencies and are more closely spaced. This demonstrates why small rooms pose mode problems while large rooms are mainly unaffected.

Graphing Room Modes

There are several ways to depict the variations of modal sound pressures throughout a specific space. The drawings in Fig. 5-9(A) are based on vectors of sound pressure rising up from the floor to a height depending on the sound pressure at that point.

Mode Number	Integers p q r	Mode Frequency, Hz	Axial	Tangential	Oblique
1	1 0 0	23.0	x		
2	0 1 0	35.3	x		
3	1 1 0	42.7		x	
4	2 0 0	48.6	x		
5	0 0 1	56.5	x		
6	2 1 0	60.1		x	
7	1 0 1	61.4		x	
8	0 1 1	66.6		x	
9	0 2 0	70.6	x		
10	1 1 1	70.8			x
11	3 0 0	72.8	x		
12	2 0 1	74.5		x	
13	1 2 0	74.5		x	
14	3 1 0	80.9		x	
15	2 1 1	82.5			x
16	2 2 0	85.7		x	
17	0 2 1	90.4		x	
18	3 0 1	92.2		x	
19	1 2 1	93.5			x
20	4 0 0	97.0	x		
21	3 1 1	98.7			x
22	3 2 0	101.4		x	
23	2 2 1	102.6			x
24	0 3 0	106.0	x		
25	0 0 2	113.0	x		
26	3 2 1	116.1			x
27	2 3 0	116.6		x	
28	0 1 2	118.4		x	
29	0 3 1	120.1		x	
30	1 1 2	120.8			x
31	1 3 1	122.5			x
32	2 0 2	123.0		x	
33	2 1 2	128.0			x

TABLE 5-2 Mode Calculations for a Large Rectangular Room (Room Dimensions: 23 ft, 4 in × 16 ft × 10 ft)

Mode Number	Integers p q r	Mode Frequency, Hz	Axial	Tangential	Oblique
34	3 3 0	128.6		x	
35	2 3 1	129.6			x
36	0 2 2	133.2		x	
37	3 0 2	134.4		x	
38	1 2 2	135.4			x
39	0 4 0	141.3	x		
40	2 2 2	141.8			x
41	4 3 0	143.7		x	
42	3 2 2	151.8			x
43	0 3 2	154.9		x	
44	1 3 2	156.8			x
45	3 4 0	158.9		x	
46	2 3 2	162.4			x
47	0 0 3	169.5	x		
48	1 0 3	171.2		x	
49	3 3 2	171.2			x
50	0 1 3	173.1		x	
51	1 1 3	174.8			x
52	2 0 3	176.3		x	
53	2 1 3	179.8			x
54	0 2 3	183.6		x	
55	3 0 3	184.5		x	
56	1 2 3	185.2			x
57	2 2 3	189.9			x
58	4 0 3	195.3		x	
59	3 2 3	197.5			x
60	0 3 3	199.9		x	
61	2 3 3	205.7			x
62	0 4 3	220.6		x	
63	0 0 4	226.0	x		
64	3 0 4	237.4		x	
65	0 3 4	249.6		x	

TABLE 5-2   (*Continued*)

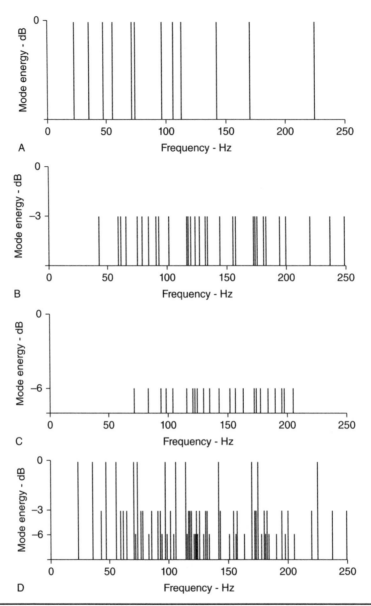

Figure 5-10 Plot of modes for the large room specified in Table 5-2. (A) Axial. (B) Tangential. (C) Oblique. (D) Combined.

In (1, 0, 0), there is a plane of zero pressure at the center of the room running from front to back and floor to ceiling. Maximum sound pressure (1.0 unit) prevails over the entire surface of the left and right ends of the room. The floor plan of Fig. 5-9(B) is a flat-plane contour plot of sound pressure. Each of the contour lines is the bottom edge of a plane running from floor to ceiling. For example, the 0.5 pressure line shows that 50% pressure prevails everywhere on the plane from floor to ceiling.

In Fig. 5-9(C), the pressure graph is moved to one side. Such a graph could appear on all four edges of the floor plan if desired. These graphs in (C) indicate the sound pressure on and above the floor area. These same (A), (B), and (C) ideas are carried on through the (2, 0, 0) and the (3, 0, 0) modes, using the same principles to convey comparable sound pressure information.

Experiments with Modes

Simple experiments can illustrate the variations in frequency response with respect to location that are created by standing waves in a room. A loudspeaker, driven by an oscillator/amplifier combination, fills the room with sound, playing a frequency. For example, in a room that is 20 ft in length, some of the resonant frequencies associated with the dimension will be 28 Hz, 56 Hz, 85 Hz, and 113 Hz. These are calculated by dividing the speed of sound by the wavelength of the resonant frequency, which is twice the room dimension [$1130/(2 \times 20) = 28$, etc.]. Listeners walking the length of the room will hear significant variations in intensity; it helps to plug one ear so that only one location is heard at a time. The fluctuations in intensity result from the vector summation of many modal resonance components. Alternatively, listeners can stay stationary in the room as the oscillator is slowly swept through the low-frequency range. Even if the output of the oscillator, amplifier, and loudspeaker are constant, the listener will hear variations in amplitude at different frequencies as modes associated with those frequencies are excited and the stationary listener finds him or herself positioned at different places along each of those standing waves. This variation in sound level, according to frequency and location, is what makes room modes problematic.

Modal Distribution in Nonrectangular Rooms

Some room designers encourage the use of splayed walls and nonrectangular spaces to improve the distribution of sound. In the adaptation of an existing space, the splaying of walls is usually prohibitively expensive. In new construction, wall splaying can often be achieved at modest increase of cost.

There is no question as to the effectiveness of wall splaying in the elimination of flutter echo. Flutter echo occurs when the space between two parallel reflecting surfaces is excited by some impulsive sound. The train of multiple reflections with its characteristic fluttering sound can be very audible and disturbing. Splaying walls to eliminate flutter echo, however, is excessive. Flutter echo is easily eliminated by the strategic placement of a few patches of absorber or diffuser.

The goal of breaking up standing wave patterns in the space is another reason advanced for the use of nonrectangular rooms. In Fig. 5-11, the rectangular room is indicated by broken lines, and the skewed room by solid lines.

In going from a rectangular room to its skewed counterpart, as shown in the figure, the null of a (1, 0, 0) mode is only distorted, not eliminated. Likewise, the normal-mode frequencies are all changed slightly, not eliminated. In some designs, the heavy, outer walls of control rooms of recording studios are splayed. Whether this practice is justified is controversial. Some would prefer the predictability of the standing-wave sound field of a rectangular room over the intractable, distorted field of a skewed room.

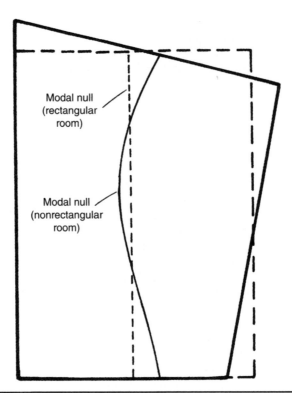

Figure 5-11   Effect of wall splaying on room mode distribution.

Modal Width and Spacing

Modal resonances separated too much in frequency from their neighbors can create problems. Ideally, the response curve of a room can be treated as a vectorial superposition of all the individual resonance curves. If these resonances are spaced greatly, the room response curve will be irregular. Separation is not the only problem; damping, excitation, and phase of individual modes also affect their contribution to the overall superimposed curve in any given situation. This helps explain why small rooms often have irregular bass response. The small dimensions place the low-frequency modes far apart in frequency so individual modes, or gaps between them, can boost or cut the frequency response. In larger rooms, modes are more closely spaced, so individual modes play less of a role; the room's response is determined by the continuum of many modes.

As noted, Fig. 5-8 plots the three mode types from Table 5-1, as well as the combined response, for frequencies below 250 Hz. The spacing between these modes, particularly adjacent axial modes, is a matter of importance. These axial modes tend to dominate the acoustics of the room. If one axial mode is separated too far from its adjacent neighbors, it tends to act alone. Ideally, the coupling of adjacent modes is desirable for smooth room response. As noted at the end of this chapter, it is difficult to correlate room mode spacing with sound quality because modes cause a room's response to be very different in different locations; a listener in any one location cannot equally hear all the modes.

The modal frequencies of Figs. 5-8 and 5-10 are depicted as narrow lines. Actually, each mode is a resonance curve with a small, but finite, width. The bandwidth of a mode is the width of the resonance curve in hertz at points 3 dB below the peak response. The bandwidth of these curves is determined by the absorption in the room. The width of a mode can be estimated by the statement $2.2/\mathrm{RT}_{60}$, where RT_{60} is the reverberation time of the room. For example, if the reverberation time of a studio is 0.5 sec, the modal bandwidth is about 4.4 Hz.

In the example of Fig. 5-8, the average spacing is 3.3 Hz. If the bandwidth of each mode is 4.4 Hz, it would appear that in a good percentage of the cases, due to overlapping of the skirts of the curves, the curves produce some coupling of adjacent modes. With a reverberation time of 1.5 sec (such as an auditorium), the modal bandwidth would be about 1.5 Hz. These would be sharp tuning curves.

Axial modes tend to dominate the response, but tangential and oblique modes of lesser amplitude tend to "fill in" between the axial modes. In Table 5-3, the axial modes of Table 5-1 below 300 Hz have been separated out and arranged in a form convenient for analyzing the acoustical properties of this room. With dimensions of 12 ft, 5-1/2 in × 11 ft, 5 in × 7 ft, 11 in, this room is quite small. In the right-hand part of Table 5-3, the axial-mode resonances are arranged in ascending order and the differences between adjacent axial modes are listed. Several of these separations are greater than 25 Hz, which should serve as a warning of possible sound coloration in the 98 to 136 Hz and 148 to 181 Hz regions.

	Length $L = 12.5$ ft $f_1 = 565/L$	Width $W = 11.4$ ft $f_1 = 565/W$	Height $H = 7.9$ ft $f_1 = 565/H$	Arranged in Ascending Order	Difference
f_1	45.3 Hz	49.5 Hz	71.5 Hz	45.3 Hz	4.2 Hz
f_2	90.7	98.9	143.0	49.5	22.0
f_3	136.0	148.4	214.6	71.5	19.2
f_4	181.4	197.9	286.1	90.7	8.2
f_5	226.7	247.4	357.6	98.9	37.1
f_6	272.1	296.8		136.0	7.0
f_7	317.4	346.3		143.0	5.4
				148.4	33.0
				181.4	16.5
				197.9	16.7
				214.6	12.1
				226.7	20.7
				247.4	24.7
				272.1	14.0
				286.1	10.7
				296.8	

TABLE 5-3 Axial-Mode Resonant Frequencies of a "Non optimum" Room (Room dimensions: 12 ft, 5.5 in × 11 ft, 5 in × 7 ft, 11 in)

The 24.7 Hz separation between 247 and 272 Hz would probably not cause colorations of speech because it is close to 300 Hz, the upper limit.

Modal resonances that are too close together can also create irregularities in frequency response. This is especially true when two or more modes are coincident, and this coincidence has mode gaps on either side. Further, let us assume that two adjoining resonances are excited by a signal. When the signal changes in spectrum and/or amplitude, the excited modes are left to decay on their own. A mode excited by a pure tone will, when the excitation is removed, decay smoothly. Under the same circumstances, two modes close together beat with each other during the decay, resulting in an irregular decay curve. Irregular decays can be audible.

Early studies of sound in small studios of the British Broadcasting Corporation revealed that voice colorations (distortion) resulted from axial modes or groups of axial modes that were separated by more than 25 Hz (Gilford, 1972). An amplifier was devised by which a narrow band of the spectrum could be amplified to any degree required. A voice signal that was noticeably colored (by the acoustics of the space) was passed through this amplifier. This peak of amplification was applied to the voice signal and adjusted until the coloration was clearly identified, and its frequency noted. A modal resonance irregularity was invariably identified with the voice coloration. Gilford suggests that coloration is audible when an axial mode is separated from its neighbors by about 20 Hz or more. Colorations appear chiefly in the region of 75 to 200 Hz with a subsidiary peak around 250 to 300 Hz. Colorations are rare below 80 Hz because of the low energy content of voice signals in that region. Colorations disappear above 300 Hz because of the increased number of all modes at higher frequencies. A contrasting theory attributes voice coloration to reflections from untreated walls, observation windows, or corner reflections. These early reflections are only slightly attenuated and form a comb-filter spectrum with a definite pitch that may explain the coloration of sound. According to this study, because of speech colorations, axial-mode separations greater than about 25 Hz are viewed with suspicion.

"Optimal" Room Proportions

If all three dimensions of a room are the same (the room is a cube), all three axial-mode frequencies would coincide, as well as their multiples. The same would be true of the tangential and oblique modes. It is obvious that by making the three major dimensions of a room different, the cubical problem would be avoided and the modal frequencies would be better distributed. The question is, what dimension ratios are preferred?

Choosing room dimension ratios for "optimum" distribution of modal frequencies was a favorite study in earlier days. A selection from the work of Sepmeyer is given in Table 5-4 (Sepmeyer, 1965). His mathematical analysis indicated that the room dimension ratios A, B, and C provide more optimum distribution of modal frequencies. Table 5-5 lists the modal frequencies for a room following Sepmeyer's C ratio.

The small room analyzed in Tables 5-1 and 5-3, and Fig. 5-8 is quite non optimum, as the ratios of dimensions do not come close to those of Table 5-4. Studying the difference column in Table 5-5 shows that the room in this example, following the C set of ratios of Table 5-4, has smaller differences than the non optimum room in Table 5-3. The procedure followed in the room designs in the later part of this book uses the ratios of Table 5-4. While some dimension ratios should clearly be avoided, there is no optimal set of ratios.

	Height	Width	Length	
A	1.00	1.14	1.39	
B	1.00	1.28	1.54	
C	1.00	1.60	2.33	
Example: Assume Height of 10 ft.				**Volume**
A	10.00 ft	11.4 ft	13.9 ft	1743 ft³
B	10.00	12.8	15.4	2035
C	10.00	16.0	23.3	3728

TABLE 5-4   Rectangular Room Dimension Ratios for Favorable Distribution of Modal Frequencies (*Sepmeyer, 1965*)

	Length $L = 23.3$ ft $f_1 = 565/L$	Width $W = 16$ ft $f_1 = 565/W$	Height $H = 10$ ft $f_1 = 565/H$	Arranged in Ascending Order	Difference
f_1	24.2 Hz	35.3 Hz	56.5 Hz	24.2 Hz	11.1 Hz
f_2	48.5	70.6	113.0	35.3	13.2
f_3	72.7	105.9	169.5	48.5	8.0
f_4	97.0	141.3	226.0	56.5	14.1
f_5	121.2	176.6	282.5	70.6	2.1
f_6	145.5	211.9	339.0	72.7	24.3
f_7	169.7	247.2		97.0	8.9
f_8	194.0	282.5		105.9	7.1
f_9	218.2	317.8		113.0	8.2
f_{10}	242.5			121.2	20.1
f_{11}	266.7			141.3	4.2
f_{12}	291.0			145.5	24.0
f_{13}	315.2			169.5	0.2
				169.7	6.9
				176.6	17.4
				194.0	17.9
				211.9	6.3
				218.2	7.8
				226.0	16.5
				242.5	4.7
				247.2	19.5
				266.7	15.8
				282.5	0.0
				282.5	8.5
				291.0	

TABLE 5-5   Axial-Mode Frequencies of an "Optimum" Room Using Sepmeyer's C Ratio (Room Dimensions: 23 ft, 4 in × 16 ft × 10 ft)

As described below, the entire quest for optimal ratios is problematic. Indeed, in most rooms, some treatment will always be needed to smooth low-frequency response.

Practical Limitations

Clearly, room modes have adverse effects on the low-frequency response of a room and are particularly troublesome in small rooms. It seems logical that rooms should be carefully designed for optimal dimensions. For example, as we have seen, room dimensions that are equal or simple multiples of each other should be avoided. Researchers have identified room-dimension ratios that more evenly distribute modal frequencies across the spectrum. When possible, this countermeasure should be implemented.

However, as Toole points out, it is difficult or impossible to correlate optimal room dimensions with perceived sound quality (Toole, 2008). Methods to describe optimal dimensions assume that all modes will be equally excited and that a listener will equally hear all modes. Unless the sound source is located in a corner, and the listener is in an opposite corner, these assumptions do not hold. Moreover, the methods may wrongly assume that the three types of modes are equal in intensity and that most rooms are rectangular with flat, reflective surfaces. As a result, in nonideal (real-world) rooms with the listener in a nonideal position and with multiple sound sources in nonideal locations, the predictions of the optimization models cannot be realized. A better approach may be one in which specific listener and multiple-subwoofer locations are tailored for given room dimensions (Welti and Devantier, 2006).

In most cases, realistically, although precautions can be taken, room mode low-frequency issues will exist in most small rooms. Thus, because they can never be eliminated in the design phase of a new room, room modes must be identified and treated with acoustical measures. For example, the drywall construction found in many buildings will provide broadband low-frequency absorption that will tend to flatten the room's response curve. By decreasing bass energy, there is less constructive interference to produce peaks, and less destructive interference to produce nulls. Other measures such as bass traps and Helmholtz resonators can also be used to address problem modes. With careful attention, modes can be managed and small rooms can exhibit very satisfactory low-frequency response.

CHAPTER 6

Sound Isolation
and Site Selection

The ear is an adaptable organ. After working under noisy conditions for long periods of time, one grows accustomed to the noise. Remember the lighthouse keeper who, when the foghorn suddenly stopped, was startled by the silence and exclaimed, "What was that?" Microphones and amplifiers have no such adapting ability, and will always truthfully register noise as noise. Thus a low noise level is an important goal for an acoustically sensitive room, and one requiring considerable effort to achieve.

Any noise that passes through isolating barriers is recorded along with the desired signal. Likewise, noise can disrupt a listening session. Some noises are more annoying than others. For example, consider a soft musical passage as the signal. Against that signal, the sound of a dog barking would be especially prominent, while that of a low-level continuous hiss would be much less noticeable. Such reasoning is not of much comfort. Our goal must be to reduce all environmental (and electronic) noises to an absolute practical minimum.

How low must the noise level be within an acoustically sensitive space? We may also ask, if noise standards are relaxed, how serious are the consequences? The answers depend on how critical the recording and listening will be. A specification for a low interior noise level can be difficult and expensive to achieve. The answers to the questions are also thus highly influenced by budget.

Site selection is supremely important for determining the quality of the interior acoustical space, and the cost to build it. It is a truism that the quieter the location of a proposed studio, the easier it will be to achieve the required low noise level within it. The opposite of this is also a truism, and a very expensive one: The noisier the location, the heavier must be the walls, floors, and ceilings, and the relatively greater the degrading effect of weak links such as windows, doors, and their seals. A low background noise level within a recording studio or other audio space is difficult and expensive to achieve structurally; it is imperative that much thought be given to site location.

Sound Isolation

The principle of sound isolation can be demonstrated with a simple electric buzzer experiment suggested by Egan (Egan, 1972). In Fig. 6-1 (A), a sound level of 70 dB is measured with a sound-level meter as a buzzer rests on the table. In Fig. 6-1 (B), a box made of porous glass fiber is then placed over the buzzer. The sound level is still 70 dB inside the

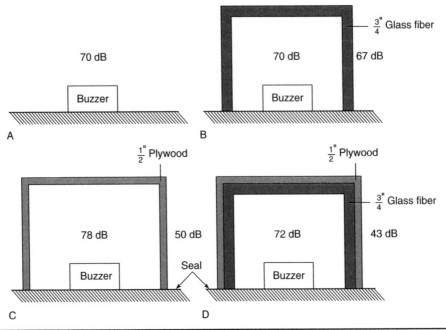

Figure 6-1 A sound isolation experiment showing the benefit of an airtight solid barrier with lining. (A) Buzzer in open. (B) Buzzer covered with box of 3/4-in glass fiber. (C) Buzzer covered with box of 1/2-in plywood. (D) Buzzer covered with plywood box lined with glass fiber.

glass-fiber box, and the outside level is reduced only slightly to 67 dB. In Fig. 6-1(C), a plywood box with an airtight seal to the table is placed over the buzzer. The inside sound level is now 78 dB, but the level outside has been reduced to 50 dB. In Fig. 6-1(D), the glass-fiber box is slipped into the plywood box and the combination placed over the buzzer and sealed to the table. The sound level inside the box is now 72 dB, and the level outside is 43 dB.

The noise reductions in the four cases are tabulated in Table 6-1; they are: (A) 0 dB, (B) 70 – 67 = 3 dB, (C) 78 – 50 = 28 dB, and (D) 72 – 43 = 29 dB. In Fig. 6-1(B), the glass-fiber box offers only a tiny reduction of noise energy as it passes through the light-weight glass fiber. In Fig. 6-1(C), the plywood box noise reduction of 28 dB is due to the much higher mass of the plywood as compared to that of the glass fiber in Fig. 6-1(B). The sound is trapped within the plywood box of (C) by multiple reflections, and the reverberant noise level within is increased from 70 to 78 dB. A somewhat greater noise reduction results when the plywood box is lined with glass fiber in Fig. 6-1(D). Lining the box with glass fiber reduces the reverberant noise level within to 72 dB, giving a slight increase in noise reduction (to 29 dB, which is only 1 dB better than the plywood box alone). However, since the noise level at the buzzer is reduced by 6 dB, the noise level outside the box is substantially reduced to 43 dB.

We observe that absorbents such as glass fiber absorb sound, but they are ineffective in blocking sound and instead allow sound to pass through them. The absorbents are low in density compared to the plywood and, for this reason, are quite ineffective as

Condition	Sound Level Near Buzzer (dB)	Sound Level Outside (dB)	Noise Reduction (dB)
Buzzer in open	70	70	0
Buzzer covered with box of 3/4-in glass fiber	70	67	3
Buzzer covered with box of 1/2-in plywood	78	50	28
Buzzer covered with plywood box lined with 3/4-in glass fiber	72	43	29

TABLE 6-1 Sound Isolation Experiment

sound barriers. Density (mass) is needed for effective noise reduction. For this reason, for example, any suggestion of reducing the noise from a neighbor's apartment by covering the shared wall with carpet or other absorbent should be rejected as futile.

When evaluating noise attenuation, it is always important to remember the logarithmic nature of decibel measurements. For example, a manufacturer might boast that a partition provides a 50% decrease in noise level. However, this is a decrease of only 3 dB, which is barely noticeable to most listeners. Likewise, a 75% decrease is 6 dB, and a 99% decrease is 20 dB. For this reason, because dB measurements more closely correlate to how we hear sounds, dB measurements should be used when describing sound isolation. Also, more specialized descriptive methods such as transmission loss and sound transmission class are widely used.

Transmission Loss

When sound passes through a barrier such as a window or wall, some sound energy is lost; the sound energy on the other side of the barrier is lower. The reduction in sound energy caused by the barrier is called the transmission loss (TL). Transmission loss is a property of a barrier; it is measured in decibels. To achieve good sound isolation, a higher TL is desirable. Massive barriers such as brick walls offer a high TL while lightweight barriers such as most window glass have a low TL. The transmission loss of a material varies with frequency. In some cases, isolation is expressed as τ, which is the ratio of sound energy transmitted to the incident sound energy. TL (in dB) can be found: TL = 10log $1/\tau$. When τ is 1 (such as for an open window), TL is zero; τ is a dimensionless quantity; that is, units are not used.

Transmission loss values are measured in a laboratory and the values are assigned to a material as part of its specification. These measurements are performed under carefully controlled conditions; for example, materials are secured in resilient mounts. The measured results represent a best-case scenario. In practical construction conditions, transmission loss performance will often be somewhat less (perhaps 10 dB). For example, sound leaks and flanking paths around the barrier will reduce TL. This is discussed in more detail later.

Insulation versus Isolation

The distinction between insulation and isolation should be clarified. A given wall construction, for example, will offer a certain transmission loss to sound traveling through it. The absorptive sound transmission loss can properly be referred to as insulation. A studio under construction will have many sound transmission loss figures applying to it, for the wall, the floor, the ceiling, the observation window, and so on. The term insulation can be applied to the sound transmission loss of each element.

Isolation is more a general term applying to structures as a whole. For example, the difference between the sound-pressure levels taken inside and outside a studio would be a measure of the isolation of the studio. High isolation is the design goal; numerous attenuations measured as sound transmission losses (insulations) is the way isolation is achieved. In many texts, the two terms are used interchangeably.

Sound Transmission Class (STC)

The Sound Transmission Class standard rates the sound-blocking ability of a material or structure. The STC is a single-number rating of airborne transmission loss measured at sixteen 1/3-octave bands with center frequencies ranging from 125 to 4000 Hz. The higher the STC value, the greater the sound insulation provided. The practical benefit of higher STC ratings, expressed in terms of sound intrusion between rooms, is shown in Table 6-2.

The STC is a practical simplification for the sake of convenience. The measurement of the transmission loss of a partition, graphically presented as a function of frequency, provides an accurate way to describe the partition's noise-blocking efficiency. An example of the measured transmission loss (of a wall construction) is shown in Fig. 6-2. The standard STC contour given for the wall construction is shown as a broken line. The STC is described in *ASTM Standard E413-87 Classification for Rating Sound Insulation*. STC values are based on decibels, but they are dimensionless (no units) quantities.

The shape of the standard STC contour is described by the data in Table 6-3. To use the STC contour, it is first plotted to the same frequency and transmission-loss scales used in the transmission-loss graph. A vertical line at 500 Hz should be included. This plot is then laid over the transmission loss curve, aligning the 500-Hz line of the overlay with that of the graph. The overlay is then shifted vertically with both 500-Hz lines together until the following conditions are met:

1. The maximum deviation of the test curve below the contour at any single test frequency shall not exceed 8 dB.

STC Range	Intrusion between Rooms
0–20	Voices clearly heard between rooms
20–40	Voices heard in low background noise
40–55	Only raised voices heard in low background noise
55–65	Only high-level noise heard in low background noise
70	Practical limit

TABLE 6-2   Sound Intrusion between Rooms According to STC Ratings

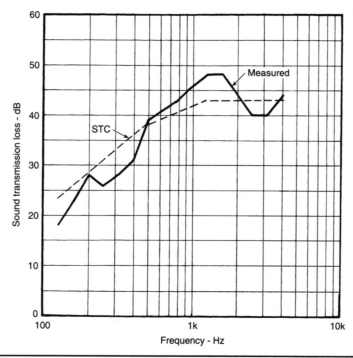

FIGURE 6-2 STC plot of a wall showing both the measured performance, and its standard quoted performance.

Frequency* (Hz)	Sound Transmission Loss (dB)
125	24
160	27
200	30
250	33
315	36
400	39
500	40
630	41
800	42
1000	43
1250	44
1600	44
2000	44
2500	44
3150	44
4000	44

*1/3-octave intervals.
(*ASTM E413-87*)

TABLE 6-3 Standard STC Contour

2. The sum of all the dB of the transmission loss curve variations at all 16 frequencies of the test curve below the contour shall not exceed 32 dB (an average deviation of 2 dB).

When the position of the STC contour meets these two conditions, the STC value of the partition can be read at the 500-Hz ordinate. The STC contour is expedient because it gives a single-number rating of the partition's ability to block sound that conforms reasonably well to practice. For this reason, STC is widely accepted. With this short-hand method of evaluating partitions, it is easy to compare one structure with another with regard to their ability to block sound. For example, a barrier with an STC-38 rating can be compared to another barrier with an STC-49 rating, and there is some sense in saying one offers 11 dB more transmission loss than the other. However, the frequency response of the STC curve must also be considered.

Room Noise Reduction

In some cases, the isolation between two rooms is referred to as the noise reduction (NR). The noise reduction is the difference between the sound levels in two rooms at a specified frequency. For example, if the sound level in a noisy room is 90 dB at 500 Hz, and the level is 55 dB in an adjoining room, the NR of the partition is 35 dB. Note that the NR of the partition does not depend on the levels themselves; in this example, no matter what level is tested, the wall's NR remains 35 dB.

Noise reduction is similar to transmission loss and values for NR and TL are usually within a few decibels of each other. However, TL is a rating of a barrier that is determined by laboratory testing; in some ways, TL is a prediction of ideal performance. NR is a practical figure measured in rooms; for example, the size of the partition, the absorptivity in the receiving room, and sound leaks can affect NR.

The Mass Law

When a sound wave strikes a barrier such as a wall, its energy causes the wall to move back and forth. This physical movement is small, but it is enough to cause the wall to reradiate the sound wave on the other side of the wall. Thus some of the sound energy can pass through a wall. Generally, the heavier the wall, that is, the greater its mass, the less a sound wave will cause it to move. Thus less energy can be reradiated on the other side. Particularly at lower frequencies, heavier partitions are more effective at stopping sound than lighter partitions. As noted below, other factors such as the coincidence effect also play a role in the effectiveness of partition isolation.

The mass law gives the average transmission loss for a diffuse source of sound as a function of the wall surface weight and frequency. The surface weight is the density (or weight in pounds) of one square foot of partition surface. The mass law provides a convenient approximation to the performance of single partitions. Considering only the mass of a partition, the sound transmission loss increases about 5-dB for each doubling of its surface weight. For a perfectly limp panel (one without any structural stiffness), this figure is 6 dB. A normal partition has some stiffness, hence the 5-dB figure is closer to actuality. For example, consider three concrete slabs with thicknesses of 3, 6, and 12 in with surface weights of 40, 80, and 160 lb/ft^2, respectively. Their STC values might be 40, 45, and 50. Clearly, the law of diminishing returns applies; each 5-dB

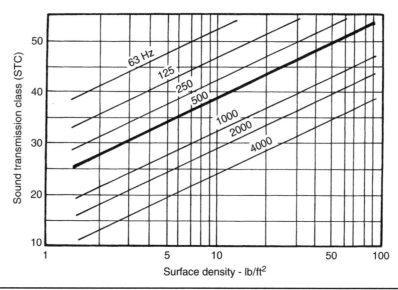

Figure 6-3 STC rating versus surface density, shown at different frequencies. An STC rating at 500 Hz is commonly used as a reference.

increase in STC requires a doubling of the homogeneous surface weight. This soon becomes impractical; to overcome this, more sophisticated partition designs are used. In particular, multilayered partitions using different materials are employed; the different densities of the partitions, the separation between them, and the decoupling between partitions all combine to yield more effective attenuation.

Actual transmission losses vary with surface density and frequency as shown in Fig. 6-3. Transmission loss increases sharply with density and frequency. The 500-Hz line is made heavier to emphasize that it is commonly used to compare partitions as a midscale, midband approximate representation of the entire frequency effect.

The Coincidence Effect

An examination of many sound transmission loss curves often shows a pronounced dip in the curve in the midrange region. For example, let us assume the dip occurs around 2500 Hz. We are concerned about this dip because the transmission loss decreases where this dip occurs. This means that noise can more easily penetrate the barrier in that vicinity. This dip in transmission loss can be 10 to 15 dB, and depends on the type of material and its properties such as bending stiffness and damping. For example, stiff materials have a greater coincidence effect dip. That is, less stiff materials are preferred.

This dip is associated with the bending vibration of the partition. The partition vibrates at its natural frequency of resonance when it is excited by sound energy of that frequency. In Fig. 6-4, the compression wavefronts (solid lines) coincide with the positive peaks of wall vibration, and the rarefaction wavefronts (broken lines) coincide with the negative peaks at this resonant frequency. Due to the vibration of the partition, some sound is radiated on the other side of the partition. Thus the sound transmission loss is decreased at or near this resonant frequency. This coincidence effect occurs when the wavelength of the

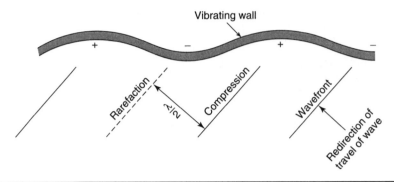

FIGURE 6-4 A graphical demonstration of coincidence dip.

incident sound coincides with the wavelength of the resonant frequency of the partition. In our example, the wavelength of 2500-Hz sound is about 0.45 ft.

In addition to the wall's back-and-forth motion, which depends on its weight or mass, a wall can have other motions. Other resonances depend upon a wall's bending stiffness. The practical effect of all these resonances shows up in the transmission loss curve. The coincidence effect is especially interesting because of its prominence on transmission loss curves.

Lead-Loaded Vinyl

The ideal isolation material has the properties of being both dense and limp. Sheet lead clearly meets these criteria and for this reason its transmission loss follows the prediction of the mass loss. Moreover, lead can be easily folded, formed, cut, and crimped. With its mass, even a relatively thin sheet provides good isolation and since it is not stiff, there is little dip from the coincidence effect. In tight places where many materials may be impossible to use, lead sheet can be fitted around pipes, ducts, and other uneven shapes to ensure that all potential sound leaks are plugged. Sheet lead is available in vinyl or other kinds of wrapping and is specified in lbs/ft^2. Alternatively, and usually at a lower cost, other types of mass-loaded vinyl sheets are available for acoustical isolation purposes.

Sound Leaks and Flanking

Sound energy travels readily through air. When designing and constructing a sound-isolating partition, it is very important to make sure the partition is airtight. Any opening in the partition, however small, will allow sound to easily travel through the partition. Looked at in another way, an air leak has a TL of 0 dB. An opening of just 0.01% in a wall surface might reduce the wall's overall transmission loss by 10 dB or more. The more urgent the need for sound isolation, and the higher the intended TL of the partition, the greater the relative damage caused by a sound leak.

In practice, this means that care must be taken during construction to caulk and seal all holes and cracks and any other potential air leak. Otherwise, the isolating properties of even the most massive partition will be compromised. Porous absorbers such as glass fiber should not be used to plug leaks; rather, nonhardening flexible caulk is preferred. All potential weak links must be avoided; for example, the entire perimeter of a stud and gypsum-board partition should be caulked. Also, the perimeters of doors must be tightly gasketed and clearly, openings such as in louvered doors must be avoided.

When constructing walls, AC outlet or other service boxes should not be placed back-to-back on opposite sides of a wall. Rather, the boxes should be spaced apart with at least one stud between them; also, any openings in the boxes should be sealed. Connecting conduit should be internally sealed as well. Fixture mounting, bad and good, is shown in Fig. 6-5. When possible, to help avoid sound leaks, fixtures should be mounted on the wall surface instead of using in-wall mounting.

Flanking occurs when sound finds a path around a barrier. For example, a plenum is the open area above walls that do not reach the structural ceiling. Clearly, this gap poses a problem. Sound can easily pass from one room to another via the plenum. Any gaps above walls must be sealed. Similarly, any ductwork in the plenum passing between rooms must be isolated. Otherwise, even if there are no vents in the rooms, sound will enter the ductwork in a noisy room and easily travel through the duct to other rooms. To avoid this, the ducts in the noisy room should be isolated, for example, in drywall tunnels. When vents are present in the ducts, baffles or more elaborate measures must be taken; these are described in more detail in Chap. 10. As another example, an attic space or crawl space that is common to several rooms can create opportunities for flanking paths.

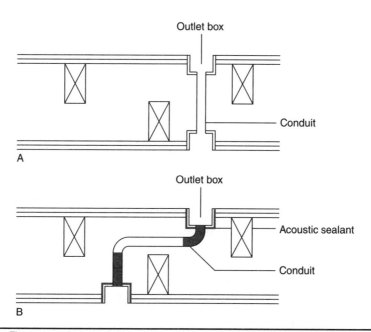

Figure 6-5 Fixtures such as outlet boxes must be mounted so as to preserve the isolation of a wall. (A) Outlet boxes are mounted back-to-back; there is no acoustic sealant; the conduits are not sealed. This construction will compromise the isolation of this double staggered-stud wall. (B) Outlet boxes are spaced apart; acoustic sealant is used; the conduits are sealed. This construction helps maintain the isolation of the wall.

Other Potential Problems

To achieve a low noise level in a room, every aspect of the design and construction must be carefully scrutinized. That is because almost every part of a room design can create a noise problem. For example, it can be difficult to isolate plumbing noises. Resilient mounts must be used to decouple pipes and fittings from wood or metal framing as well as gypsum-board walls. Otherwise, water noises can be easily transmitted.

Great care must be taken during construction; design or construction errors must be eliminated. It is critical to maintain expert supervision and employ workers who are experienced in building acoustically sensitive spaces. A good acoustician will use a stethoscope to check for isolation defects in walls, windows, and doors. Furthermore, periodic checks should be performed to ensure that isolation has not deteriorated over time. For example, hardening caulking or misalignment of seals created as a building settles can decrease isolation that was originally present.

Checklist of Building Materials

Subject to building codes, and the architectural design itself, a variety of common materials can be used in the construction of a building. Each material will provide different amounts of absorption and insulation with a unique frequency-response characteristic.

Whenever materials are being considered, it is important to remember that every material affects a room's acoustics, often in ways that are hard to anticipate. For example, if a room has a wood parquet floor and a plaster ceiling, a potential problem of flutter echoes can be addressed by carpeting the floor. However, the carpet will add considerable high-frequency absorption to the room, and therefore the room may have relatively too little low-frequency absorption; this may create a bass-heavy sound. To address this, panel absorbers may be needed. Thus solving one problem creates another, which must also be solved. Further analysis may show that the panels create yet another problem; for example, perhaps the panels cause unwanted reflections, which in turn must be addressed. Experienced acousticians think through all the consequences of room treatment and devise solutions that can solve a problem without creating another problem, or even solutions that can solve multiple problems. Let us now consider some typical building materials.

Masonry. Poured concrete of normal weight provides considerable attenuation to airborne sound, but will transmit impact and structureborne noise, and provides almost no absorption. Brick and other types of masonry behave similarly with high attenuation of airborne sound but provide little absorption. Stone material can provide significant attenuation, and is highly reflective. Normal-weight concrete masonry units (such as cinder blocks) provide good attenuation to airborne sound particularly if the hollow cells are filled with sand; its slightly porous surface provides small amounts of absorption but this can be diminished if the units are coated with a sealing paint.

Glass. Single panes of glass provide little attenuation to airborne sound. Glass provides almost no absorption at high frequencies, but can act as a resonator and provide absorption at lower frequencies. Some types of laminated glass provide increased sound attenuation compared to sheet glass of the same thickness.

Gypsum Board. Gypsum board is a lightweight material but it can provide a fair amount of attenuation, especially when multiple layers are used and carefully sealed.

Gypsum board does not provide absorption at high frequencies, but when used with an airspace, it can act as a resonator and provide low-frequency absorption. Lightweight steel studs used for partition framing offer slight resiliency, and thus decouple partition faces and improve attenuation. Heavy steel studs and wood studs offer less resiliency, and hence less decoupling and less attenuation; attenuation can be improved by adding porous absorption in the interior cavity and by using resilient channels. When a partition is fabricated with two independent rows of studs supporting two independent surfaces, the decoupling provided by stud resiliency is not necessary and any kind of stud can be used.

Plywood and Paneling. Plywood does not provide significant attenuation, and offers little absorption at high frequencies; when used as a resonator, it can provide significant low-frequency absorption. Similar to plywood, wood paneling on furring adds little attenuation to a wall but can absorb considerable low-frequency energy.

Floor Coverings. Carpet can be effective for treating both airborne and structure-borne noise. Carpet provides good sound absorption at higher frequencies; it is one of the few floor coverings to do this. In most carpeted rooms, the carpet provides the majority of high-frequency absorption. Absorption is improved when carpet is thick and is laid over porous (open cell) padding. Carpet is also extremely useful for attenuating high-frequency impact noise on concrete slab floors; however, on wood floors, low-frequency impact noise can remain problematic. There is evidence that absorption coefficients for carpets are often not accurate; when possible, absorption from the carpet should be measured directly. The downside to carpet is that while it efficiently absorbs high frequencies, it absorbs almost no low frequencies. A room with extensive carpeting will likely require targeted low-frequency absorbers to provide a balanced sound. Otherwise, the room may have a bass-heavy, "boomy" sound. Resilient tile (such as vinyl) provides little attenuation or absorption; however, it can decrease impact sounds by cushioning the impact.

Curtains. Heavy curtains can provide absorption while thinner curtains are far less effective. The curtains should be draped with deep folds or hung away from the rigid surface, and must not be applied directly to the wall surface. Acoustically transparent curtains and other materials can act as porous facings for glass fiber or other absorbing panels.

Glass Fiber. Glass fiber is widely used as a low-cost sound-insulating and thermal-insulating material; it is an efficient sound absorber. Glass fiber (also known as fiberglass) is available in boards, blankets, and batts; it is also used for resilient support as compressed blocks and sheets. In general, the thicker the glass-fiber material, the greater is the absorption. Glass-fiber material can be placed behind an acoustically transparent fabric or grille or manufactured as wall or ceiling tiles; in either case, mid- and high-frequency absorption is quite high. Glass fiber can provide additional attenuation when loosely placed inside a partition such as a stud wall or loosely placed in a plenum.

Specialized Products. A variety of specialized building products are available; they are designed to provide specific acoustical characteristics. Acoustical foam (such as polyurethane) manufactured with an open-cell structure provides poor sound isolation, but good absorption. Absorption increases with thickness of the foam and airspace behind the foam. A common plaster skin has attenuation and absorption characteristics similar to gypsum board. Acoustical plaster is more porous than common plaster and provides only a fair amount of absorption. Several types of proprietary acoustical plaster are available. Acoustical tile (such as cellulose or mineral fiber or more efficiently, glass fiber, ceiling tiles) provides good absorption; low-frequency absorption is improved when tiles are suspended from a grid. Some tiles, particularly mineral fiber tiles with a foil backing, also provide attenuation and can be employed below plenums. As noted above, lead sheet or leaded vinyl sheets provide a high degree of attenuation. Because of its pliability, lead sheet can be used to seal breaches in irregular spaces such as in plenums. Duct lining, usually of glass fiber, provides good absorption and noise attenuation along an interior duct length. Attenuation is best at mid and high frequencies. Many types of nonhardening compounds are used to seal cracks, joints, and wall penetrations; an airtight seal is critical to prevent sound leaks, which compromise the attenuation properties of a partition. Mineral or cellulose fiber sprays are used for thermal insulation and fire-proofing; when applied with a thick coat, they can provide absorption. Coustone, also called quietstone, is an example of a sturdy, washable porous absorber. It is formulated from bonded flint aggregate, is rigid and heavy, and presents a hard-wearing surface.

Locating a Studio

Of all the acoustically sensitive spaces, a recording studio is perhaps the most critical in terms of sound isolation. In a venue such as a concert hall, a fleeting sound may cause a minor annoyance; in a recording studio, the same sound could become part of a recorded track. Either the track would have to be rerecorded, or else the sound would be a permanent fixture that would provide a legacy of annoyances. In practical terms, a high-quality studio demands an ambient noise level that is as quiet as possible.

In some locations, background noise levels are extremely low. For example, in wilderness areas of some national parks, the $L_{90(1)}$ reading (discussed in Chap. 11) might be below 10 dBA in octave bands from 250 to 2000 Hz. Unfortunately, it is rarely possible to find a building site with ambient noise levels so low. In contrast, a quiet suburban neighborhood at night, in the winter, might have noise levels measuring 30 to 40 dBA. Noise surveys are often used to evaluate ambient sound levels at a site; these are discussed in Chap. 11.

At the other end of the scale, according to the Occupational Safety and Health Administration (OSHA), an instantaneous peak sound level of 140 dB or higher has the potential to cause immediate and permanent hearing loss. Much lower levels can cause noise-induced hearing loss over long-term (decades) exposure; for example, levels of 70 to 85 dB can be a concern. However, sound levels much lower than these, sounds that pose no health threat, can be highly detrimental to the success of many rooms.

Recording studios and other acoustically sensitive spaces are rarely located in isolated buildings with no exterior noise problems. In most cases, the building is uncomfortably near a road, flight path, or other noise source. For example, vehicular traffic running through an "urban canyon" can generate unacceptably high noise levels as sound reflects from one building surface to another and balconies reflect sound back onto the street level. In addition, it may be necessary to deal with other building tenants, some of whom might be serious noisemakers in their own right. For example, an offset printing shop, machine shop, carpenter shop with power tools, or an automobile repair facility can all pose serious problems. Such circumstances require special solutions and can significantly increase design and construction costs. The cost of locating a building away from noise sources must be compared to the cost of providing the building with sufficient isolation to be able to operate in a noisy environment. In particular, it can be very expensive to isolate a building in an environment with low-frequency noise. In that case, seeking another and quieter site may be more cost-effective than providing acoustical isolation.

The best way to achieve a quiet space is to locate the building away from any external noise sources. Roadways and highways, airport flight paths, railroad tracks, heavy-industry factories, and similar noise sources will necessitate rigorous isolation treatment. Choosing a site away from these sources will lower costs. For example, a recording studio located at least 3 miles from a commercial airport can be designed with a much lower isolation requirement. In some cases, a site in an otherwise noisy area might be suitable because other buildings block sound from noise sources such as highways. When choosing a quiet site, as much as possible, anticipate future building projects that may bring unwanted noise to the area. Also consider that otherwise favorable zoning assignments can be changed in the future.

Many noise sources are transient, and are thus more difficult to anticipate. For example, the heavy machinery that is part of any city landscape can appear outside any street address, and possibly stay for days, weeks, or months. Whenever possible, an acoustically sensitive facility, particularly one that is a commercial business, should be designed to isolate the interior from unexpected transient noise sources. Table 6-4 lists examples of heavy equipment and the maximum noise levels they can create when operating.

Noise Source	Noise Level (dBA)
Augured earth drill	94 at 10 ft
Backhoe	87–99 at 30 ft
Backhoe, idling	74 at 30 ft
Cement mixer	77–85 at 10 ft
Chain saw cutting trees	89–95 at 10 ft
Circular saw on concrete	91 at 30 ft
Compressor, street	91 at 3 ft
Front-end loader	79–93 at 50 ft
Garbage truck	85–97 at 10 ft
Jackhammer	96 at 10 ft
Paving breaker	86 at 30 ft
Steamroller	87 at 30 ft
Street cleaner	74 at 30 ft
Street paver	84 at 30 ft
Wood chipper, tree shredding	93–101 at 30 ft
Wood chipper, idling	85 at 30 ft

(*Cowan, 1992*)

TABLE 6-4 Maximum Noise Levels from Heavy Machinery

In the home, even though most noise levels are relatively low, there are numerous interior noise sources that can interfere with home recording and listening. Outdoor light machinery such as lawn mowers and leaf blowers can generate relatively high noise levels. A barking dog might create an impulsive noise level of 90 dBA measured at 5 ft. Table 6-5 lists examples of household noise sources and the maximum noise levels they can create.

In some unique cases, people can create very loud unwanted noise levels. In particular, the spectators in sporting event venues can generate levels in excess of 100 dBA. For example, the crowd's noise level in an outdoor football stadium might reach 111 dBA and the interior of a hockey arena might reach 113 dBA. Sound levels at a school playground during recess might reach 100 dBA. Sound levels at music concerts can reach similar or higher levels.

The best way to deal with a noise problem is at the source. As one example, a recording studio in a remote area was troubled by infrequent low-frequency sounds. The noise was traced to cars passing over an old wooden bridge near the building. The most expedient solution was to pay to build a new and quieter bridge. More commonly, for example, noise sources such as HVAC units can be placed on isolation pads or moved to another part of the building.

Unfortunately, in many cases, for example, with traffic and airplane noise, source treatment is not possible. Some exterior HVAC systems need ventilation, thus isolation materials cannot be directly applied. In some cases, when the source cannot be treated, barriers can be used to mitigate noise intrusion. Barriers can successfully reduce high-frequency noise (for example, tire noise) but are much less effective with low-frequency

Noise Source	Noise Level (dBA)
Dehumidifier	58–60 at 5 ft
Food blender	76–81 at 3 ft
Garbage disposal	76–78 at 3 ft
Microwave oven	56–58 at 3 ft
Sink faucet, full force	71–73 at 3 ft
Handheld vacuum cleaner	82–87 at 3 ft
Small rechargeable vacuum cleaner	75–77 at 3 ft
Floor vacuum cleaner	78–85 at 5 ft
Steam carpet cleaner	84–92 at 5 ft
Portable hair dryer	77–86 at 1 ft
Lawn edger	89–93 at 5 ft
Leaf blower	87–93 at 5 ft
Power lawn mower, hand	81–86 at 5 ft
Power lawn mower, riding	88–93 at 5 ft

(*Cowan, 1992*)

Table 6-5   Maximum Noise Levels from Household Sources

noise (for example, engine noise). This is because the longer wavelengths of low frequencies will diffract, thus bending around obstacles such as walls. Earthen berms and masonry walls provide some isolation. The taller the barrier, the better; if possible, absorption should be placed on the barrier surface. Trees and vegetation are not particularly efficient as isolators. When using a barrier, it should be placed close to either the noise source or the desired quiet area. In addition, the barrier must extend horizontally well beyond the noise source or quiet area.

Roadway vehicle traffic poses a great challenge to acoustical designers. In an urban area, many sites will have traffic noise levels that exceed residential noise level standards. Noise from highways is particularly difficult to control. Costly remedies such as concrete barriers and earthen berms are often constructed. According to the Federal Highway Administration, very generally, a barrier will provide only 5 to 10 dB of attenuation. Because of diffraction of sound over and around barriers, their practical maximum attenuation is about 15 dB. Attenuation of at least 3 to 5 dB can be expected when the barrier interrupts the line of sight to the source. Each additional 2 ft of height adds about 1 dB of attenuation as the barrier's shadow zone increases. A barrier's length should be about four times the distance between the source and the listener. If the sound source can be seen over, through, or around a barrier, the barrier is providing no significant sound attenuation from that source.

Floor Plan Considerations

As much as possible, acoustically sensitive rooms should be located away from exterior noise sources. For example, a studio should be located on the side of the building facing away from a busy street. In addition, within a building, acoustically sensitive rooms should be separated from noisy rooms. For example, a maintenance shop should be located at the far end of a building housing a studio. Similarly, bathrooms and HVAC rooms should be located away from the studio. This can be accomplished by placing acoustically neutral rooms such as storage closets in between to act as buffers. Of course, the same considerations apply to vertical locations; for example, an HVAC room should not be placed over a studio. Clearly, windows should not face a noise source such as a busy highway.

Locating a Space within a Frame Structure

A space for a studio or other sound-sensitive activity located within a frame structure may have the thud and thump problem of footfalls to contend with (as elaborated in Chap. 8). These low-frequency components are inevitable because they are very difficult to control with lightweight frame structures. If the natural period of vibration of the floor-ceiling structure is about the same frequency as the peak energy of the thuds and thumps of footfalls, the footfall noise will be present and might actually be amplified. Carpeting stairs and hallways will reduce the higher-frequency impact components of footfalls, but the thud and thump low-frequency components will tend to penetrate floating floors, heavy carpet, or other precautionary layers. If the type of recording or playback being done is not critical, extended low-frequency response might not be required. For example, a voice-over studio is concerned only with higher frequencies. In such a case, a high-pass filter can be inserted in the signal path, which would reject both signal components and troublesome noise below about 63 Hz (see Fig. 8-2). Extreme isolation measures to eliminate these thud and thump problems in frame structures can be an exercise in futility.

Locating a Space within a Concrete Structure

Locating a studio or other sound-sensitive area within a concrete structure does offer some prospect for effective isolation against the low-frequency components of footfall thuds and thumps. Although quite expensive, it is possible to build a studio-within-a-studio offering maximum isolation from airborne and structureborne noise using familiar construction techniques such as massive partitions separated by decoupling.

In most cases, it is not cost-effective to attempt to build an ultra-quiet studio. After all, much of the world's audio and visual work is accomplished on a daily basis in environments that are less than perfect. It can be very expensive to get the last few decibels of quiet.

CHAPTER 7

Wall Construction and Performance

The design and construction of walls is fundamental to any architectural endeavor. When the space within will be acoustically sensitive, the walls take on additional importance. They must work as sound barriers to isolate the interior space from exterior noise, and to isolate the exterior from the interior sound. Before wall construction can be specified, two major questions must be answered. First, what noise level will define the environmental ambient noise in which the studio (or other sound-sensitive room) will operate? Second, what background noise level goal is to be set for the inside of the room? The difference between these two noise levels defines the transmission loss that must be supplied by the walls and other barriers.

As noted in earlier chapters, porous materials that excel as sound absorbers are poor at sound isolation. Very generally, the more massive the partition, the better the sound isolation; this is particularly true at low frequencies. Overall, sound isolation is harder to achieve at low frequencies and easier at high frequencies. But other factors influence the success of a barrier's sound isolation. For example, performance may deteriorate at mid frequencies because of a coincidence effect.

Sound can propagate through any medium; for example, it can pass through air and through solids. The former is airborne transmission and the latter can occur as structureborne transmission. For example, sound from a distant room may travel down a corridor and into your room via an air path, but it may also pass through the concrete floor that is common to both rooms and reradiate from the floor in your room. In fact, sound travels more efficiently in more dense mediums. Generally, airborne sound is higher in frequency (above 100 Hz) and structureborne sound is lower in frequency (below 100 Hz). In some cases, structureborne sound is present only as a vibration that is felt. Any barrier must be designed to minimize both airborne and structureborne transmission. Airborne transmission is minimized by sealing any air leaks in a partition; these can seriously degrade the acoustical performance of even the most formidable wall. Structureborne transmission can be reduced by decoupling elements of a partition, and thus breaking the transmission path; for example, two partition leaves separated by resilient mounts will reduce structureborne transmission. Also, structureborne sound is reduced by eliminating any resonant conditions in the transmission frequency range.

Walls as Effective Noise Barriers

A wall that is effective as a noise barrier is characterized by a high transmission loss, that is, the sound energy is significantly decreased by passing through the wall. Transmission loss varies with frequency, thus some complexity is introduced. The graphs of Figs. 7-1, 7-2, 7-3, and 7-4 show the measured transmission loss curves of four common wall structures. Such measurements require special facilities and experienced operators, and are impractical for specific structures for all but the largest jobs.

Figure 7-1 shows the measured transmission loss curve for the wall partition pictured. This wall with 3-5/8-in steel or 4-in wood studs is faced with gypsum board ("drywall" or "hardboard") of 1/2-in and 5/8-in thickness. The shape of the transmission-loss curve is anything but regular. The problem of characterizing TL curves has been simplified by the concept of the Sound Transmission Class (STC) rating, a single-number rating of airborne transmission loss. The STC contour for the partition is also shown in the figure. STC is described in Chap. 6.

The Mass Law and Wall Design

As noted in Chap. 6, sound transmission loss increases by about 5 dB for each doubling of surface mass. The surface density is the density (or weight in pounds) of one square foot of partition surface. Transmission loss increases with density and frequency.

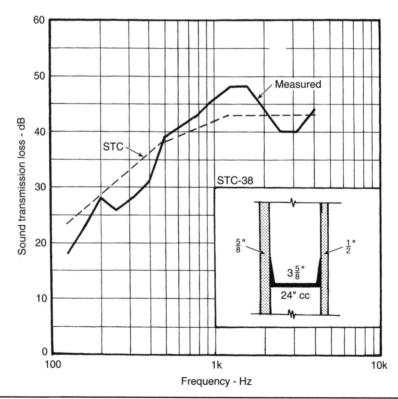

Figure 7-1 Transmission loss of a wall constructed of studs and drywall. (*Northwood, 1968*)

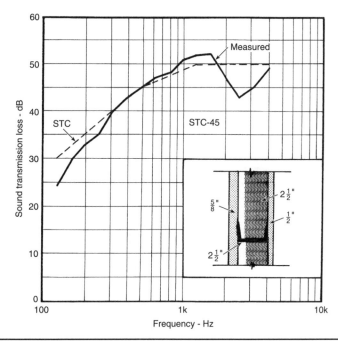

FIGURE 7-2 Transmission loss of a wall constructed of studs and drywall with glass fiber in the interior airspace. (*Northwood, 1968*)

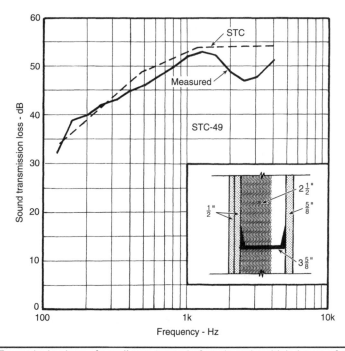

FIGURE 7-3 Transmission loss of a wall constructed of studs and multiple layers of drywall with glass fiber in the interior airspace. (*Northwood, 1968*)

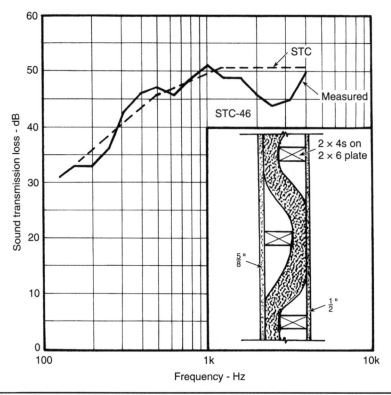

Figure 7-4 Transmission loss of a staggered-stud wall with glass fiber in the interior airspace. (*Northwood, 1968*)

Figure 7-5 illustrates a common use of the 500-Hz mass law. The black dots represent the Sound Transmission Class ratings of several partitions. Table 7-1 summarizes the data applicable to each of the seven points, pulling together the data from Figs. 7-1 through 7-4, and adding three others of a general nature for comparison.

Point #1 shows the STC 38 given by the partition of Fig. 7-1 composed of two leaves (one of 1/2-in and the other of 5/8-in drywall plasterboard), with 3-5/8-in channel and with no glass fiber in the interior space. Its surface density (4.8 lb/ft²) alone would give it a 500-Hz transmission loss of only STC 33. The actual value of STC 38 means that the two leaves separated 3-5/8 in perform better than the mass law prediction with the two leaves placed together to isolate the mass effect. In other words, the structural form has increased the transmission loss about STC 5 dB over that of the mass law alone.

Point #2 shows the performance of the partition of Fig. 7-2. Even though the surface densities of the partitions of Figs. 7-1 and 7-2 are the same (4.8 lb/ft²), the latter is rated at STC 45 while the former is only STC 38. This is in spite of the spacing of the two leaves being decreased from 3-5/8 to 2-1/2 in. The reason for this improved performance is that 2 in of glass fiber has been introduced in the interior space. Glass fiber does not increase the transmission loss directly; it is far too light in weight for that. It helps by subduing the resonances in the space, which tend to reduce

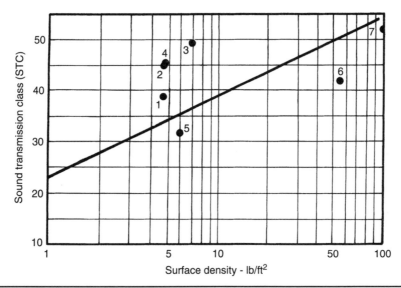

Figure 7-5 A comparison of the STC ratings of various wall structures relative to the mass-law prediction curve.

From Figure	Point on Figure 7-5	Leaf (A)	Leaf (B)	Surface Density, lb/ft²	Leaf Spacing	Glass Fiber	STC	Source
7-1	1	1/2"	5/8"	4.8	3-5/8"	—	38	N
7-2	2	1/2"	5/8"	4.8	2-1/2"	2"	45	N
7-3	3	2 × 1/2"	5/8"	7.0	3-5/8"	2-1/2"	49	N
7-4	4	1/2"	5/8"	4.8	staggered stud	yes	46	N
—	5	1/2"	1/2"	4.2	2 × 4 wood studs	—	32	—
—	6	4-1/2" brick with 1/2" plaster	—	55.0	—	—	42	—
—	7	9" brick with 1/2" plaster	—	100.0	—	—	52	—

(*N = Northwood, 1970*)

Table 7-1 Sound Transmission Loss of Partitions (Data Summary for Figure 7-5)

partition performance. Without the glass fiber within the cavity, sounds at and near the resonant frequency of the panels would pass through the wall structure with relatively little attenuation. The glass fiber adds damping to the structure, which reduces resonance. Care must be taken to add the glass fiber loosely; if tightly packed, it would tend to couple and thus bridge the panels, decreasing the transmission loss.

Point #3 shows the performance of the partition in Fig. 7-3; it indicates still more increase in the STC rating. Surface density has been increased to 7.0 lb/ft² by adding gypsum board. Spacing is 3-5/8 in, and glass fiber has been placed in the interior space. The result is an increase to STC 49, which is an increase of 15 dB over the mass law prediction.

Point #4 shows the performance for the staggered-stud wall construction of Fig. 7-4. This reaches STC 46, somewhat short of the STC 49 of Fig. 7-3 in which the surface density was greatly increased.

Point #5 shows the performance of a simple wall: 2 × 4 wood studs with 1/2-in drywall attached to each side. The performance is below that of the mass law, STC 32, for a surface density of 4.2 lb/ft². Point #2 shows that a partition close to the same surface density can give STC 45 with the addition of a heavier gypsum-board sheet and some glass fiber in the space.

Points #6 and #7 are added to show that the lightweight structures considered above can be designed to provide high STC ratings. This clearly shows the relative efficiency of more sophisticated partition designs. However, heavy walls usually provide greater insulation at low frequencies.

Wall Designs for Efficient Insulation

Figure 7-6 illustrates seven simple ways to achieve efficient insulation in a basic wall structure: (A) increase weight; (B) wider spacing of leaves; (C) staggered studs; (D) leaves of different weight; (E) resilient strips; (F) glass-fiber blankets; (G) perimeter caulking. Figure 7-6(A) demonstrates that mass is needed for a wall to be an effective barrier. This mass can easily be added to a wall in the form of sheets of gypsum board by screwing or cementing. Table 7-2 lists the surface density of common sound barrier walls, which vary between 3 and 11 lb/ft².

While mass is needed for isolation, the positioning of the mass and in particular the spacing between masses is important. If the spacing between the two leaves of a wall is eliminated or made very small, the transmission loss will be reduced to mass functioning as though it were a single-leaf wall of combined weight. For two leaves spaced widely apart and decoupled, the transmission loss of this double-wall construction approaches the sum of the two as though two separate walls existed. Our practical wall spacing of about 4 in is somewhere between the two extremes. The transmission loss is increased only a small amount by increasing the spacing from 4 to 6 in, but this is one factor in the significant increase for staggered-stud walls.

The two sides of staggered-stud walls [shown in Fig. 7-6(C)] are only coupled at the perimeters. They are normally built on 6-in plates instead of the usual 4-in plate. Though nominal in itself, the added spacing gives some increase in transmission loss. Making the mass of the two leaves different [as in Fig. 7-6(D)] makes the resonant frequencies of the two leaves different. When these frequencies coincide, the dip in the transmission loss curve is magnified at that frequency. Offsetting the resonant frequencies by making the leaf mass different tends to smooth the transmission loss curve. For example, one gypsum-board face could use 5/8-in board while the other uses a double layer of 5/8-in boards.

Figure 7-6(E) illustrates the use of resilient strips for the purpose of providing some decoupling and isolation of a layer of gypsum board from the wall itself. This also helps make the resonance points of the two leaves appear at different frequencies, avoiding deep dips in the transmission loss curve. Some resilient channels are S-shaped, with one

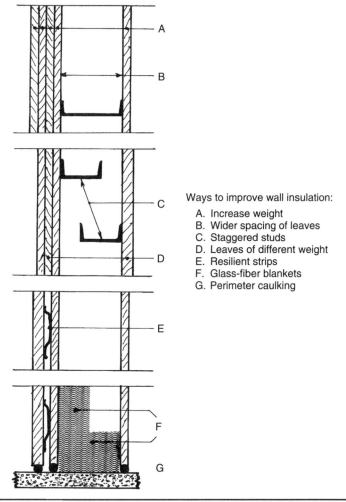

Ways to improve wall insulation:
A. Increase weight
B. Wider spacing of leaves
C. Staggered studs
D. Leaves of different weight
E. Resilient strips
F. Glass-fiber blankets
G. Perimeter caulking

FIGURE 7-6 Seven ways to improve sound insulation of a wall.

end secured to the wall and the other secured to the gypsum board. When using resilient strips, care must be taken to ensure that isolation is preserved; for example, the proper screws should be used. Also, care must be taken when mounting cabinets or other attachments to the wall; for example, screws that are too long will bridge the isolating strip by solidly attaching the gypsum board to the underlying stud.

The use of thermofiber (building insulation) in the space between the two leaves of a wall [as shown in Fig. 7-6(F)] is an economical way to increase the transmission loss of a wall by about 5 dB. This glass fiber helps to control various cavity resonances, which tend to decrease transmission loss. When low-density glass-fiber insulation is used, the cavity should be completely filled; for every doubling of insulation thickness STC increases by about 2 dB. However, glass fiber should not be packed too tightly because it could conduct sound energy through the partition. For the same reason, mineral fiber (which is stiffer) should not completely fill the cavity. Placing glass fiber

Leaf A	Leaf B	Surface Density, lb/ft^2
Unfilled Steel Stud Partition		
3/8"	3/8"	3.0
1/2"	1/2"	4.0
5/8"	5/8"	5.0
3/8" + 3/8"	3/8" + 3/8"	6.0
1/2" + 1/2"	1/2" + 1/2"	7.5
5/8" + 5/8"	5/8" + 5/8"	10.0
Unfilled Wood Stud Partition		
3/8"	3/8"	4.0
1/2"	1/2"	5.0
5/8"	5/8"	6.0
3/8" + 3/8"	3/8" + 3/8"	7.0
1/2" + 1/2"	1/2" + 1/2"	8.5
5/8" + 5/8"	5/8" + 5/8"	11.0

Note: *The densities above include weight of studs.*

TABLE 7-2 Surface Densities of Common Partitions

in the interior cavity is particularly helpful in staggered-stud walls in which the faces are acoustically decoupled from each other.

Figure 7-6(G) recalls the importance of sealing all edges of a partition. The importance of liberal use of sealant cannot be overstressed. It is an inexpensive way to achieve added transmission loss in a wall. Normal framing always results in cracks, which might not be important in nonacoustical walls, but which are very important in walls associated with studios and other sound-sensitive spaces. The cracks present in normal framing are always there, and the sound-penetrating cracks under, over, and around a wall can easily mean the loss of 10 dB of STC rating. Figure 7-7 illustrates both the problem and the solution.

For example, before the 2 × 4-in or 2 × 6-in plate is bedded on the floor, several strips of acoustical sealant should be applied to the floor and/or the plate to seal the ever-present crack between the concrete and the plate. Every edge-framing member should be so sealed. Additional sealant should be applied to the periphery of the gypsum-board wall. The sealant used should be of the nonhardening variety, often referred to as "acoustical sealant."

Improving an Existing Wall

It is often necessary to adapt an existing space for a studio or other acoustically sensitive room. Existing walls are rarely built for high transmission loss. One way to improve an acoustically weak existing wall is to construct another partition on either side of it. The design and construction of that additional wall will determine whether it succeeds as a sound barrier. Figure 7-8 details a practical approach to such a wall improvement.

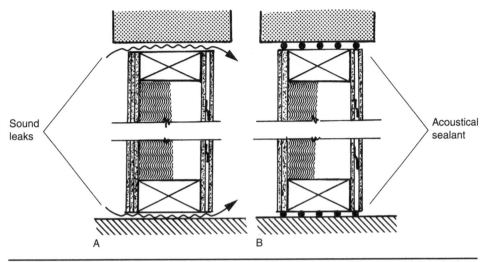

Sound leaks

Acoustical sealant

A   B

Figure 7-7   The importance of sealing wall elements. (A) Sound penetrates through air leaks in the partition. (B) Sound leaks are stopped by acoustical sealant.

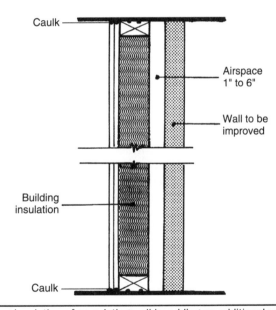

Caulk

Airspace 1" to 6"

Wall to be improved

Building insulation

Caulk

Figure 7-8   Improving insulation of an existing wall by adding an additional wall.

The airspace gap is important; 1 in is a minimum, but the gap should be increased if floor area allows. From there on, the additional wall embodies the features that have been described, such as thermofiber (building insulation) in the airspace and multiple layers of gypsum board. The outer layer of gypsum board could be mounted on metal resilient strips, and more than two layers of gypsum board could be included to yield an additional improvement in STC. Such steps may be unwarranted because of flanking sound traveling around the barrier. As another measure, a wall's transmission loss can be increased by simply adding a 1/2-in gypsum-board layer to each side of the existing wall. When possible, seals in the new layer should not coincide with seams in the original layer. This technique preserves the airspace gap between the original leaves, and adds weight to the leaves. All seams should be caulked to prevent sound leaks.

Flanking Sound

Solid materials are efficient carriers of sound. That is why you can place an ear on the rail of a railroad track and hear an oncoming train that is several miles away, but still not audible through the air. Similarly, in a high-rise concrete structure, elevator sounds and other noises can be carried throughout the structure with very little loss. Wooden structures are also good conductors of sound. Table 7-3 (Harris, 1957) compares the attenuation of sound in iron, brickwork, concrete, and wood. In concrete, for example, sound can travel 100 ft with an attenuation of only 1 to 6 dB.

Structure carries sound from one point to another, but radiation of that sound into an otherwise quiet area occurs largely because walls and floors act as diaphragms. Instead of the noise staying in the beams and columns of the structure, it vibrates the walls and floors, reradiating sound into the air within the rooms. Because of this, protection from both airborne and structureborne sound must be considered.

Figure 7-9 illustrates the travel of noise through structure to walls, which radiate into the space of a protected area. The specific case of a frame structure is shown in Fig. 7-10. Flanking sound can travel by a combination of airborne and structureborne paths from the noisy space to the protected area through an attic space or a subfloor space. Such flanking sound can easily nullify the cost and work spent on improving a shared wall.

Gypsum-Board Walls as Sound Barriers

As noted above, common wall construction with studs and gypsum board can provide relatively good sound isolation. Isolation can be greatly improved by using double layers of gypsum board; layers should be offset so that seams are staggered. When using wood studs, multiple layers should be secured with visco-elastic adhesives instead of rigid adhesives or screws. Isolation of wood-stud walls can be improved by mounting the gypsum board on resilient channels instead of directly on the studs; the gypsum board is screwed to the channels and the screws are not allowed to touch the studs.

Material	Attenuation dB/100 ft
Iron	0.3 to 1
Brickwork	0.5 to 4
Concrete	1.0 to 6
Wood	1.5 to 10

(Harris, 1957)

TABLE 7-3 Attenuation of Longitudinal Waves

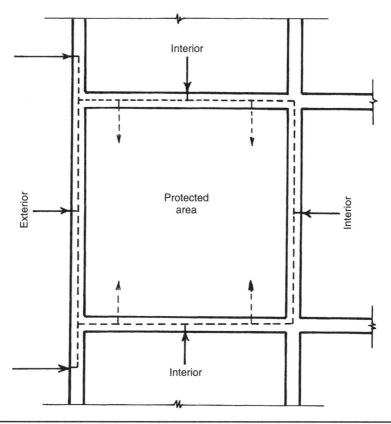

Figure 7-9 Structureborne sound intrusion.

Local building codes may dictate metal studs; because of their ability to flex, metal studs provide slightly better sound isolation compared to wood studs. With metal-stud walls, isolation is improved when gypsum board is screwed to the metal studs instead of using adhesive; a line of adhesive stiffens the flanges whereas individual screws do not. Care must be taken so that the drywall will not vibrate against the metal studs causing audible buzzes; this can be accomplished by increasing the number of screws and laying a bead of nonhardening caulk along each stud. As with any acoustical construction, any sound leaks should be sealed with caulking and flanking paths must be eliminated.

Masonry Walls as Sound Barriers

Brick and concrete block walls can provide effective sound isolation. These materials are heavy, and the mass law favors weight. Masonry walls provide relatively more isolation than lightweight walls at low frequencies; therefore, they are superior in music applications. However, the weight of concrete blocks varies, thus some walls provide better isolation than others. Isolation can be increased by adding well-tamped sand or mortar in concrete block cells. Isolation can also be increased by adding a gypsum board layer or ideally by building a double concrete block partition where the walls are

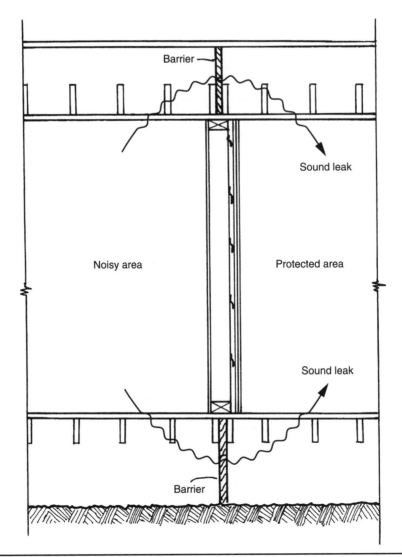

FIGURE 7-10 Flanking paths and barriers.

acoustically decoupled. Placing glass fiber or other sound absorption in wall air cavities can also increase isolation. Masonry walls do have a drawback; they are efficient transmitters of impulse noise through structureborne transmission. Therefore, care must be taken to isolate impulsive sound sources from a masonry wall.

The performance of brick and concrete block walls is summarized in Table 7-4. The STC of brick walls can be equaled by frame walls, but plastered concrete block walls, with their STC ratings of 57 and 59, are difficult to equal with frame construction. Circumstances often dictate which walls are most practical, both from acoustical and construction standpoints.

Figure 7-11 shows the smooth measured transmission loss curves of both unplastered single- and double-leaf concrete block walls. A comparison with the concrete

Wall Description	Surface Density, lb/ft²	STC	Reference
Brick wall, 4-1/2 in plastered both sides	55	42	Table 7–1
Brick wall, 9 in plastered both sides	100	52	Table 7–1
Double 6 in concrete block walls spaced 6 in	100	59	Figure 7–11
Single 12 in concrete block wall	100	51	Figure 7–11

TABLE 7-4   Summary of the Acoustical Performance of Masonry Sound Barriers

block walls of Table 7-4 shows that plastering is more effective on the single-leaf wall (an increase of 6 dB) than the double-leaf wall (no increase).

Figure 7-12 shows three examples of brick walls. An 8-in thick brick wall provides good sound isolation, but performance can be significantly improved by adding layers to one side. These layers comprise vertical wood furring strips, resilient metal channels, and 1/2-in gypsum board. Voids are filled with 2-in glass-fiber fuzz. The added layers can improve transmission loss by 20 dB or more at higher frequencies. However, a design using two 4-in brick walls separated by a 4-in airspace provides insulation that is superior to the other two designs. Sound absorption such as a glass-fiber blanket should be placed in the cavity between the two walls. This two-wall design is simple, but its total depth of 12-in is a drawback. The double wall design demonstrates the effectiveness of separation

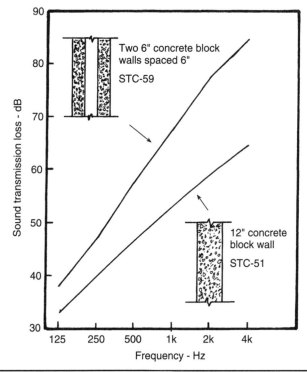

FIGURE 7-11   The transmission loss of concrete walls. (*Egan, 1972*)

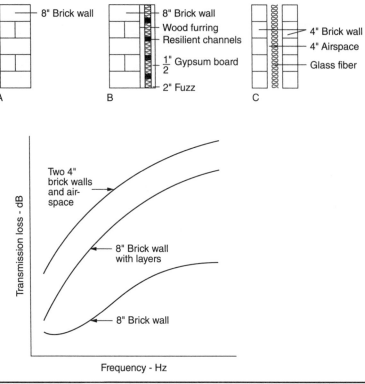

FIGURE 7-12   The transmission loss of masonry walls can be improved by adding a layer or by constructing two walls. (A) An 8-in brick wall provides good sound isolation. (B) Added layers improve transmission loss at higher frequencies. (C) Two 4-in brick walls separated by a 4-in airspace provide superior isolation.

of masses; its total weight is the same as one 8-in wall, but it widely outperforms it. For example, at high frequencies, its transmission loss may be as much as 35 dB higher.

Weak Links

The performance of sound-isolating partitions is very much affected by the weak-link principle. No matter how high the transmission loss of a barrier, it can be severely compromised by any lower-performance breach, even if that breach has a relatively small surface area. In fact, when a weak link is present, the overall isolation of the barrier will be close to that of the weak link. As noted earlier, any air leak in a partition can be very problematic. For example, Table 7-5 shows how transmission loss is degraded by air leaks of varying size in a wall with an original TL of 45 dB. For example, if the wall area is 100 ft², an air opening measuring 14.4 in² (0.1%) would decrease the wall's TL from 45 dB to 30 dB. A 1/4-in space under a door would create an opening of about this size.

The installation of electrical fixtures such as outlet boxes and studio microphone panels must be done properly, otherwise a wall's isolation can be compromised. As noted in Chap. 6, fixtures should not be mounted on exactly opposite positions (back-to-back) on a wall. Sound can leak through openings. Rather, boxes should be

Percent of Wall Area Having Air Opening	Resulting Wall TL (dB)	Resulting Decrease in TL (dB)
0.01	39	6
0.1	30	15
0.5	23	22
1	20	25
5	13	32
10	10	35
20	7	38
50	3	42
75	1	44
100	0	45

(Cowan, 2000)

TABLE 7-5 Reductions in Transmission Loss from Openings in a Wall with Initial TL of 45 dB

staggered apart; also, clearly, the boxes should be tightly caulked. When possible, electrical boxes and other services should be surface mounted.

Doors and windows can also easily undermine the acoustical integrity of a partition. For example, a brick wall might have a TL of 50 dB. When a window with a TL of 20 dB and occupying a surface area of about 10% of the wall is incorporated in the wall, the new TL may be just 30 dB. Indeed, once this breach has been introduced, the damage has been done; a much larger window area may give a TL of 25 dB.

This illustrates the difficulty of designing composite barriers. To avoid weak links, all the components of the barrier such as windows and doors must provide a transmission loss close to that of the wall itself; for example, the difference in TL should be less than 5 dB. This complicates the design and increases the costs of windows and doors. The design of windows and doors is considered in Chap. 9.

Summary of STC Ratings

Figure 7-13 summarizes the points made in this chapter regarding economical ways in which the transmission loss of a wall might be increased. Mass is the most vital element of any noise barrier. In Fig. 7-13(A), the mass of a partition is low, but it can be increased easily by adding multiple layers of gypsum-board panels. These can be screwed or cemented.

In Fig. 7-13(B), an increase in spacing of the two leaves of a partition is achieved by staggered studs. This will increase its STC value. In Fig. 7-13(C), staggered studs of metal or wood are used to increase the spacing between the two leaves. In addition, another layer of gypsum board has been added to one side to make the surface densities of the two sides different. This places their resonances at different frequencies, smoothing the transmission loss of the wall. In Fig. 7-13(D), the surface density of the two leaves is made different by using layers of wall board of different thicknesses on each side. In Fig. 7-13(E), the use of resilient strips is suggested. If one face of a wall is mounted

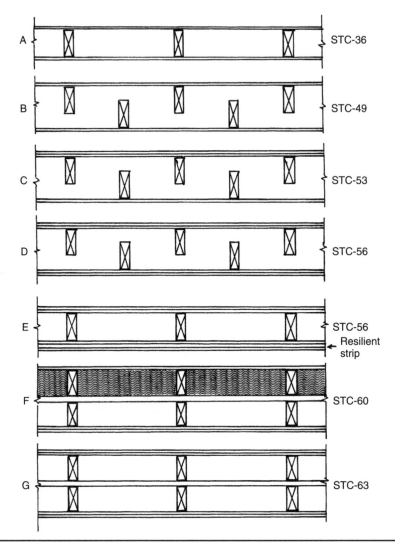

Figure 7-13 The STC ratings for various wall structures.

resiliently to the structure, the STC can be increased significantly. In Fig. 7-13(F), we see that a small improvement can be expected by using glass fiber in a cavity space, or by increasing its thickness. The improvement by adding glass fiber is modest, but is inexpensive and worth the effort. In Fig. 7-13(G), we see the use of two completely separate walls. In all arrangements, proper sealing is most important. Copious use of tubes of caulking material could be the most important single step in the entire process.

The STC rating, which attempts to give a practical and simple shortcut, has been widely accepted, but its limitations must always be kept in mind. The ultimate test of any noise barrier system, which can be made only when the studio is completed, is noise level measurements within the studio itself.

CHAPTER 8

Floor/Ceiling Construction and Performance

Floor/ceiling construction deals with many of the same issues as wall construction. In particular, the floor/ceiling must provide sufficient isolation between the rooms above and below. Clearly, noise intrusion can move in either direction, and a good floor/ceiling design can isolate a quiet lower room from a noisy upper room, or vice versa. However, unlike most wall designs, floor/ceilings must take into account the directionality of the noise transfer. In particular, it is somewhat more difficult to isolate a lower room from an upper room; this is because of footfall sounds. Footfall sounds present a challenge to builders, residents, and studio operators alike.

The sounds associated with the activity of people in the space above is a common source of complaints from the people living below. Such noises destroy the concept of privacy in living quarters and can be a source of much trouble when the noises penetrate a sound-sensitive space such as a recording studio. As with many other noise intrusion problems, the most effective remedy is stopping the noise at its source. This is the case with footfall sounds. For example, the easiest way to diminish footfall sounds (and other impact noises) in a lower apartment is to generously offer to install plush carpets and carpet pads in the apartment above.

Data on the Footfall Noise Problem

To illustrate the difficulty in designing an acoustically proficient floor/ceiling, consider the following case. Warren E. Blazier, Jr., consultant in acoustics, and Russell B. DuPree of the Office of Noise Control, State of California, gave an account of how insidious footfall noise can be (Blazier and DuPree 1994). The homeowners in a luxury condominium complex in the San Francisco area brought an $80 million class-action suit against the developers. The major claim was based on annoyance caused by footstep noise generated by the activity of residents above being transmitted through the structure to their neighbors below. Even when cooperative upstairs neighbors agreed to go barefooted or to wear soft-soled shoes, the thuds and thumps were still clearly audible below. The footfall impacts also resulted in "feelable" vibrations of the floor/ceiling structure, vibrations that even affected closet doors and light fixtures. Complaints were general throughout the building.

The average purchase price of the apartments was about $750,000, with marketing claims of acoustical privacy. Because of the upscale nature of the project, the builders

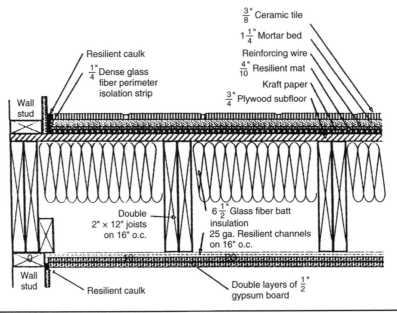

Resilient caulk

$\frac{1}{4}$" Dense glass
fiber perimeter
isolation strip

$\frac{3}{8}$" Ceramic tile

$1\frac{1}{4}$" Mortar bed

Reinforcing wire

$\frac{4}{10}$" Resilient mat

Kraft paper

$\frac{3}{4}$" Plywood subfloor

Wall
stud

Double
2" × 12" joists
on 16" o.c.

$6\frac{1}{2}$" Glass fiber batt
insulation

25 ga. Resilient channels
on 16" o.c.

Wall
stud

Resilient caulk

Double layers of $\frac{1}{2}$"
gypsum board

FIGURE 8-1 Example of floor/ceiling construction. (*Blazier and DuPree 1994*)

incorporated special design features to provide significantly better impact noise insulation than that required by California construction standards for multifamily housing. The floor/ceiling construction in question is shown in Fig. 8-1. Double joists are used to stiffen the floor plane. The ceiling of double 1/2-in gypsum board is mounted on resilient channels. The floor above starts with the usual 3/4-in plywood subfloor, on which is placed Kraft paper, a resilient mat, reinforcing wire, and 1-1/4-in mortar. On top of this is 3/8-in ceramic tile. To minimizing flanking, a dense glass fiber perimeter strip isolates the floor from the structure.

To provide data for the defense, it was decided to build an off-site laboratory mockup duplicating a pair of typical stacked rooms. The floor/ceiling structure was that described in Fig. 8-1. The lower test room was used to measure impact sound-pressure levels resulting from three types of noise sources: a standard impact tapping machine, a standardized "live-walker," and a calibrated tire drop. The measurement microphone on a 20-in boom was located in the center of the room below, and was rotated slowly in a horizontal plane during the integrating period. Data were obtained in 1/3-octave bands from 2 Hz to 4 kHz.

Among the many conclusions, the one concerning footwear is very interesting. In the live-walker tests it was found that for frequencies below about 63 Hz, the impact sound-pressure levels below the walking surface were amazingly close for leather heel/leather sole, rubber heel/leather sole, track shoe, and barefoot cases. Figure 8-2 shows that the peak energy of these live-walker tests falls in the 15- to 30-Hz region. This coincides with the fundamental natural frequency of the floor/ceiling system, which, with typical lightweight structural framing, is also between 15 and 30 Hz.

The addition of floating floors or carpeting decreases the transmission of the higher-frequency components of footfall noise, but there is no economically practical method of avoiding the thuds and thumps of footfalls with typical lightweight structural

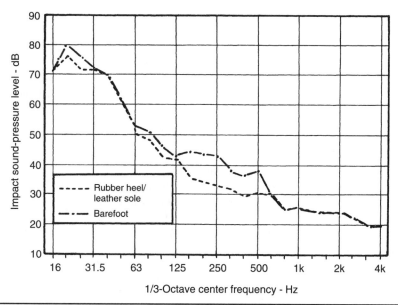

FIGURE 8-2   Measurements of floor/ceiling footfall noise.

framing. To obtain the stiffness necessary to reduce the thuds and thumps, a concrete structural floor system is required.

In Fig. 8-2, note the measurements made three octaves below the usual low-frequency limit of 125 Hz. Measuring no lower in frequency than 125 Hz misses common impact sounds 40 dB higher than at 125 Hz. Carpet is effective in reducing noises above 125 Hz, but completely ineffective in the 15- to 30-Hz range. Noise in the frequency range below 30 Hz is a concern even when footfall noise is not an issue; for example, many audio play-back systems with subwoofers can easily reproduce music in this region.

Floor/Ceiling Structures and Their IIC Performance

As noted, floor/ceilings are particularly prone to impact noise. Walking with hard heels, dropping objects, moving chairs, and similar activities can create noise that is radiated downward through the floor to the room below. Impact noise can also radiate outward through structural elements to adjacent rooms. Impact noise is most effectively treated at the source. Hard-surface floors such as concrete and terrazzo perform poorly. Soft floor coverings such as carpet with pad and rubber tile perform much better. Floating floors offer superior performance. Ceilings can be improved to some degree by adding suspended tiles. When very high isolation is required, several or all of these measures may be needed.

Impact Insulation Class (IIC) is a single-number rating of the impact sound performance of floor/ceiling constructions over a standard frequency range. This impulse single-number rating is comparable to the STC single-number rating for steady-state sounds. The higher the IIC rating, the more effective the floor/ceiling construction in reducing transmission of impact sounds such as footfalls. Numerically, IIC might vary from 20 (poor performance) to 60 (good performance). IIC data is measured at sixteen 1/3-octave bands from 100 to 3150 Hz and a reference contour is used to yield the IIC value. The curve-matching method for determining IIC values is similar to the method used for determining STC ratings. Note that an IIC rating does not consider noise below a 100-Hz center frequency; a floor/ceiling could have a high IIC value, but still transmit considerable low-frequency noise. IIC is described in the *ASTM Standard E989-89 Standard Classification for Determination of Impact Insulation Class (IIC)*.

Frame Buildings

The first and most distressing factor in floor/ceiling problems is that often there is no access to the floor side—the side that is most amenable to remedies. Instead, the only hope for improving the insulation of a room from the neighbors' noise from above is what can be done to the ceiling. Floor/ceiling constructions in frame structures are quite limited in scope, and most of them would be included in the three designs shown in Fig. 8-3.

In Fig. 8-3(A), the 2 × 10-in joists commonly have a 5/8-in gypsum-board ceiling nailed to the lower edges and a 1/2-in plywood subfloor above. The finish flooring above is probably something very much like the 25/32-in tongue-and-groove oak flooring pictured. If this is the "as found" condition, what can be done without access to the floor side?

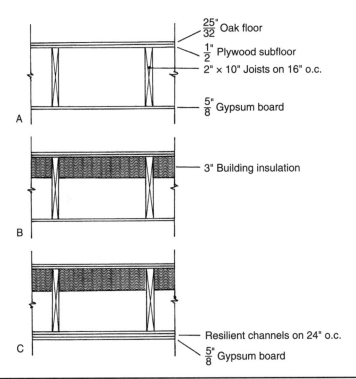

Figure 8-3 Three common floor/ceiling constructions. (A) The "as found" construction. (B) Glass fiber is added. (C) Glass fiber and resiliently mounted gypsum-board layers are added.

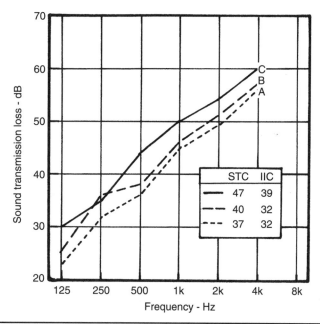

FIGURE 8-4 Transmission loss comparison of the three floor/ceiling constructions shown in Fig. 8-3. (*Egan, 1972*)

The first step would be to remove the gypsum-board ceiling fastened to the lower edge of the joists and place 3-in glass-fiber insulation between the joists as shown in Fig. 8-3(B). This insulation could be 6-in or 10-in thick, but very little improvement would be gained by going beyond the 3-in batts.

The ceiling itself can be improved by mounting resilient strips on the joists and a layer of gypsum board to the strips, and then adding a second layer of gypsum board as shown in Fig. 8-3(C). The resilient strips could instead be mounted between the two layers of gypsum board. We are still limited by not having access to the floor side above. What performance can be expected from these improvements made to the ceiling?

Figure 8-4 gives the transmission loss performance of the three floor/ceiling arrangements of Fig. 8-3. Note that the curves are quite smooth and free from untoward resonance effects. The as-found condition (A) gives STC-37, and adding the glass fiber (B) gains only 3 dB, yielding STC-40. Improving the ceiling (C) by adding two layers of gypsum board with one of them resiliently mounted brings the rating to STC-47, a gain of 7 dB. From the as-found situation of Fig. 8-4(A) a total of 10 dB in STC rating has been gained by the addition of glass fiber and resilient channels. This is close to the maximum improvement that can be expected. Figure 8-4 also includes IIC measurements.

Resilient Hangers

As noted, resilient channels can be used to provide some insulation to gypsum-board sheets. Another system that can replace or add to the resilient channels is illustrated in principle in Fig. 8-5. If some ceiling height can be sacrificed, a suspended ceiling can be

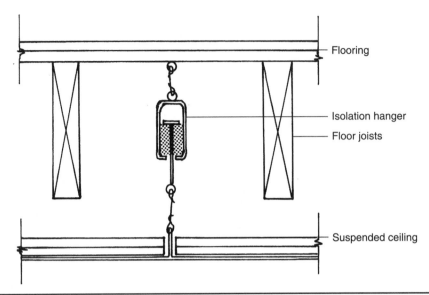

Figure 8-5 An example of a resilient hanger. The hanger is selected so that its deflection falls within its specified range.

installed, one that is isolated from the structure by resilient hangers. The resilience of the hanger illustrated in Fig. 8-5 is provided by the shaded material, which could be compressed glass fiber of the proper density. The load of the ceiling gypsum boards and supporting frame must be distributed between hangers to keep the deflection of each hanger within its rated deflection range.

Several hanger designs are available. The two hangers shown in Fig. 8-6 utilize neoprene alone or a combination of neoprene and a steel spring. The frequency ranges of these two products differ; the selection of the proper hanger for the specific job is important, and data about each is available from the manufacturer. The weight of the ceiling must be carefully matched to the deflections of the hangers to achieve maximum insulation. The periphery of the suspended ceiling should be sealed with a nonhardening acoustical sealant.

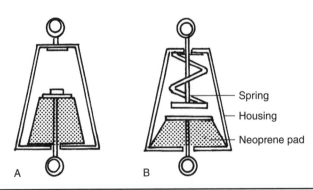

Figure 8-6 Other forms of resilient hangers. (A) Neoprene pad. (B) Neoprene pad and steel spring.

Floating Floors

In almost all cases, the best way to control noise is to address the issue at the source. In the case of a floor/ceiling problem, therefore, treating the floor of the room above will yield the most effective results. Specifically, if you have access to the room above, perhaps with the cooperation of the persons there, a floating floor could possibly be installed.

Floating floors are normally not used in budget construction. They are more typically found in more expensive concrete structures. However, floating floors can be designed for frame structures too. The budget floating floor shown in Fig. 8-7 might be appropriate in some applications.

The weight of the two layers of gypsum board and the finish floor is supported by a layer of glass fiber. The springiness of the partially compressed glass fiber is an essential component of its design. If it is totally compressed, the plywood/finish-floor layers would essentially rest on the basic floor, and little sound insulation would result. An important principle is that the gypsum-board/finish-floor mass must not touch the structure, including the edges. In Fig. 8-7, the glass fiber turned up at the edges is used to isolate the floor mass edge from the structure. A strip of compressed glass fiber could also be used to provide insulation from the wall.

A concrete floating floor can be built by placing compressed glass-fiber cubes, molded neoprene cubes, or other isolation devices across the structural floor; the cutoff frequency of the system must be calculated. Compressed glass fiber board is placed around the perimeter of the room to isolate the floating floor from the structure; a wooden strip is placed on top of the board. Plywood sheets are laid on the mounts and the edges are secured with metal straps. A pouring form is created by placing a plastic vapor barrier over the plywood sheets and up the perimeter boards. Reinforcing mesh is placed within the center of the floating slab.

To ensure the performance of the floating floor, when the floating slab is poured, concrete from the pour must not leak down to touch the structural floor and the slab must not touch the structural elements of the room. After the slab is cured, the plastic sheet can be cut and the wood perimeter strip removed. The resulting perimeter gap

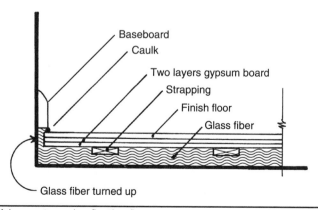

FIGURE 8-7   A low-cost wooden floating floor.

should be sealed with a nonhardening sealant. In some designs, the isolation pads have metal housings and an internal screw. The floating slab is poured directly onto the vapor barrier resting on the structural slab, and after it is cured, the floating slab is raised using floor jacks from above. When constructing any floor, particularly a concrete floating floor, it is important to anticipate placement of conduits and other infrastructure that may lie beneath the floor.

Attenuation by Concrete Layers

The improvement in attenuation that might be expected by adding a 1-1/2-in layer of troweled cellular concrete of density 105 to 120 lb/ft³ is shown in Fig. 8-8 (Grantham and Heebink, 1973). The two floor/ceiling constructions in the figure are identical, except one has 3-1/2 in of glass fiber and resilient channels and the other does not. An STC-59 rating is attained by the construction having the glass fiber and the resilient channels, a very worthwhile increase of 10 dB. This 10-dB increase in STC compares to the 10-dB increase from STC-37 to STC-47 of Fig. 8-4 obtained by adding some gypsum board, glass fiber, and resilient channels to the bare-bones "as found" condition of Fig. 8-3 (A).

Adding a concrete topping, more gypsum-board layers, glass fiber, and resilient channels increases a poor floor/ceiling structure with an STC-37 rating to a very

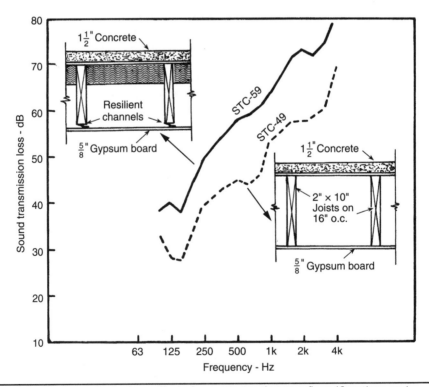

Figure 8-8   The transmission-loss effect of concrete topping on a floor. (*Grantham and Heebink, 1973*)

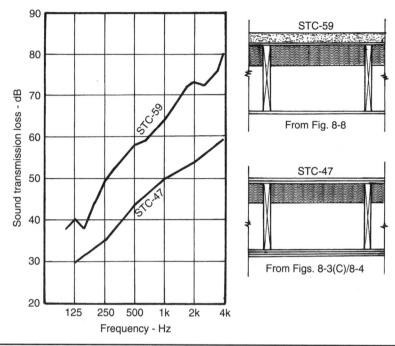

Figure 8-9 An example of a concrete layer that significantly improves the transmission loss.

respectable STC-59 rating. This is a substantial improvement in STC that will be meaningful in the finished structure.

Figure 8-9 shows a comparison of the best floor/ceiling structure without a concrete layer with the best floor/ceiling structure with a concrete layer. An estimate of the effect of the concrete layer at different frequencies can be made by comparing the STC-59 curve of Fig. 8-9 with the STC-47 curve below it. The concrete contributes about 10 dB at low frequencies, about 15 dB at 500 Hz, and about 20 dB above 2 kHz.

Plywood Web versus Solid-Wood Joists

Some floor/ceiling systems in frame structures use plywood web beams instead of conventional 2 × 10-in or 2 × 12-in wood joists. They are made with 2 × 3-in flanges and a 3/8-in plywood web. These plywood web beams, illustrated in Fig. 8-10, offer certain advantages. As they are usually fabricated off the building site, they contribute to construction speed and efficiency. From the standpoint of the architect or designer, 12-in plywood web beams allow greater spans. From the standpoint of the acoustician, they contribute to the stiffness of the floor plane and thus provide higher transmission loss to low-frequency noises.

Figure 8-11 presents the transmission loss characteristics of two good floor/ceiling systems, both offering an STC-58 rating (Grantham and Heebink, 1973). Both use resilient channels, both have 3-1/2-in glass fiber, and both have a 3/4-in plywood subfloor. The floor/ceilings differ only in that one uses 2 × 12-in plywood web joists while the

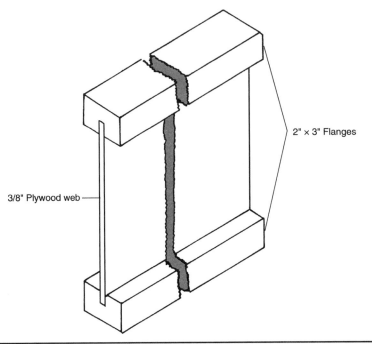

FIGURE 8-10   An example of a plywood web beam that can be used as a floor/ceiling joist.

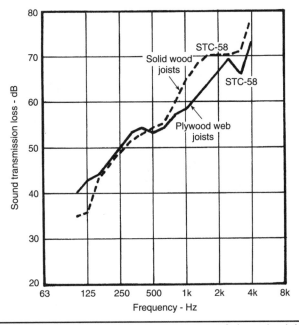

FIGURE 8-11   A comparison of the transmission loss response of plywood web joists versus solid-wood joists. (*Grantham and Heebink, 1973*)

other uses 2 × 10-in solid wood joists. There is little consistent difference between the two curves. The plywood web joists give 5-dB-higher attenuation to sound at 100 Hz than the solid wood joists. The stiffening of the floor plane by the plywood web joists results in greater low-frequency attenuation, which is noticeable in Fig. 8-11. The reverse is true in the 1000 to 2000-Hz region, where the solid wood joists offer greater attenuation than the plywood web joists. Carpets help in the 1000 to 2000-Hz region, but not in the 15 to 31-Hz region.

CHAPTER 9

Window and Door Construction and Performance

S ometimes it is important to see what is going on in another room. For example, a recording engineer in the control room must know what musicians are doing in the studio. The reverse is also often true. Glass establishes visual communication between two rooms and is far preferable to a video link. However, if a glass window is set in a high-transmission-loss wall, the transmission loss of the window must ideally meet the same specifications as the wall. This presents a challenge.

There are many kinds of glass. Over 98% of the glass produced in the United States is made by the float glass process. The molten glass is poured continuously from a furnace onto a large bed of molten tin. The molten glass literally floats on the surface of the tin, slowly solidifying as it travels over the tin. After several hundred feet of travel through a lehr, it emerges as a continuous layer of glass at approximately room temperature. The glass now is flat, fire finished, and has virtually parallel surfaces.

Float glass provides high transparency and low distortion for many applications including use in observation windows for studios and other sound-sensitive rooms. Particularly when relatively thick panes, and multiple panes, are used, float glass can provide adequate sound insulation. In some designs, laminated glass is used to provide higher levels of insulation. As in any sound-sensitive design, care must be taken during construction to ensure that the window is built without leaks or other detriments.

Although a room can be constructed without windows, it cannot function unless it has doors. Moreover, whereas many windows are designed to be permanently sealed, useful doors are designed to be opened. This basic requirement of doors makes it more difficult to achieve a reliable seal against sound leaks. Also, door thresholds are subject to wear and tear from foot traffic as well as moving equipment; this makes it difficult to provide floor seals that are durable over time.

A good door design thus requires both a door panel that is sufficiently massive and solid, and also a closure system that reliably seals the entire perimeter around the door. A good door can be constructed from common building materials, but in many cases, specially engineered acoustical doors are selected. The latter can be quite expensive.

Finally, when a door is open, no matter what its cost, it provides absolutely no acoustical insulation. To overcome this, a sound lock can be incorporated into a room design. The double doors in a sound lock add additional insulation when both doors are closed and at least some insulation when one door is open. Sound locks remove some of the burden from individual door design, but are an option only when sufficient floor space is available.

The Observation Window

Some studios located in highly scenic spots incorporate a large view window. Windows such as these can make it difficult to achieve sufficiently low background noise levels, and can create unwanted interior reflections. If the acoustically difficult idea of view windows can be successfully rejected in the early planning stages, the only significant glass left in a recording studio is the observation window between the control room and the studio. Occasionally an observation window is placed between the studio and an exterior hall.

An observation window is important in a studio's work flow, but it is also acoustically challenging. The partition between the studio and the control room must provide high transmission loss so that high sound-pressure levels in the studio do not interfere with control-room monitoring. Similarly, high-level monitoring must not intrude into the microphones in the studio. An observation window may comprise a large percentage of the partition area, and must offer very high transmission loss that is comparable to that of a massive wall. Many precautions are taken to achieve this. For example, multiple panes are used, the panes are thick, and the panes are carefully mounted in the partition.

The Single-Pane Window

The single-pane (also called single-leaf or single-glazed) window, such as the common household window, provides relatively poor sound insulation. Those living near a highway or an airport, for example, find that windows of this type allow significant noise to pass through. The mass of the glass pane controls the passage of sound through it according to the expression:

$$TL = 20 \log (fm) - 48 \tag{9-1}$$

where TL = transmission loss, dB
$\quad f$ = frequency, Hz
$\quad m$ = surface density, lb/ft^2

Taking the density of glass as 160 lb/ft^3, the surface density may readily be found for any thickness of the glass pane. The value is also often provided by the manufacturer. There is some lack of unanimity regarding the value of the constant to be subtracted. Quirt recommends 48 dB, but others prefer 34 dB (Quirt, 1982). This is of oblique importance in this discussion, because the transmission loss of various glass arrangements to be presented has been determined by measurements. The mass law, expressed graphically in Fig. 9-1, emphasizes that mass is the major component of transmission loss of glass. Transmission loss increases as the thickness increases. Also, clearly, transmission loss increases with frequency. Different glass-pane configurations (multiple panes, differing thickness, and subduing of resonances) will be presented as ways to further enhance the transmission loss of an observation window.

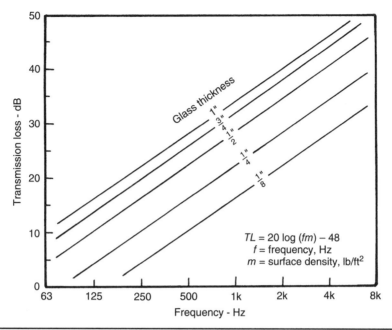

FIGURE 9-1   The mass law applied to glass panes showing panes of different thicknesses.

Window panes should be mounted with neoprene perimeter gaskets; this is pre-ferred over caulk or putty. Sealed windows will provide 3- to 5-dB greater isolation than comparable windows that can be opened, even when the openings are gasketed. When opening windows are required for example, for ventilation, double-pane, double-sash windows are preferred over single-pane, single-sash windows.

The Double-Pane Window

If a single pane offers insufficient transmission loss, the logical conclusion is that more panes will offer greater loss. There are many qualifications to this statement, but the transmission loss of the double-pane (also called double-leaf or double-glazed) windows of Fig. 9-2 (Sabine, 1975) shows an increase in transmission loss over the single-pane mass law of Fig. 9-1. Generally, a double-pane window (two panes separated by an air-space) can provide about 30 dB of transmission loss.

In Fig. 9-2, two double-pane windows with approximately 4-in spacing are com-pared on the basis of thickness of glass. The solid curve (Libby-Owens-Ford) represents measurements made with a 1/8-in pane and a 3/32-in pane. The broken-line curve (National Bureau of Standards, 1975) shows measurements made on one pane of 1/2 in and a second pane of 1/4 in. The reasons for using glass of dissimilar thicknesses will be discussed later. The thinner glass window and the heavier glass window have similar transmission loss in the 1- to 3-kHz frequency region, but the heavier glass is far superior below 1 kHz. The Sound Transmission Class (STC) ratings, which show only an advan-tage of 4 dB for the heavier glass, are quite inadequate for describing the performance of these two windows over the audible band.

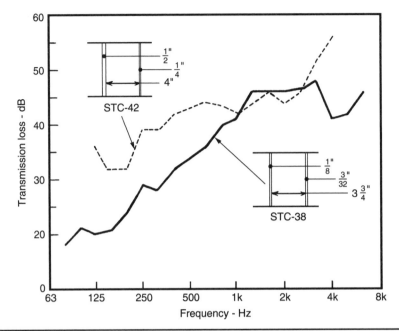

FIGURE 9-2   A comparison of transmission losses in two double-pane windows. Solid line: 1/8-in pane and 3/32-in pane. Broken line: 1/2-in pane and 1/4-in pane. (*Sabine, 1975*)

The measured transmission loss of another double-pane window is shown in Fig. 9-3 (Quirt, 1983). In this window, two 1/4-in glass panes are separated by 6 in. The overall transmission loss of the 6-in spacing is quite similar to that of the heavier glass window of Fig. 9-2. The only way the modest specific effects of dissimilar panes, glass surface density, and glass spacing can be observed is in direct comparisons such as will be made later. The irregularities of the usual measured transmission-loss curves, due principally to resonances, tend to hide these other variables of direct interest.

Acoustical Holes

An "acoustical hole" is a phenomenological name for a region of the audible spectrum in which sound more readily passes through a wall or window. In other words, sound within a narrow frequency range is attenuated several decibels less than at other frequencies. The hole appears on the transmission-loss curve as a dip at the frequency of the hole. Acoustical holes are usually traceable to resonances; this is illustrated in the examples below.

Several different resonance effects alter the shape (and effectiveness) of the transmission-loss curve of any glass window arrangement. The mass-air-mass resonance of a double window is the result of the mass of one glass pane being coupled to the other glass pane by the air in the cavity between them. Sound impinging on the first glass pane causes it to vibrate, the air in the cavity acts like a spring, and this causes the second glass pane to vibrate. This resonant system can be likened to masses attached to each end of a spring.

Quirt studied the effect of glass spacing on two 1/8-in glass panes (Quirt, 1982). He found that at 250 Hz the sound transmission-loss curve went through a very pronounced null at spacings between 1/2 in and 1 in, as shown in Fig. 9-4. At 800 Hz, the sound transmission

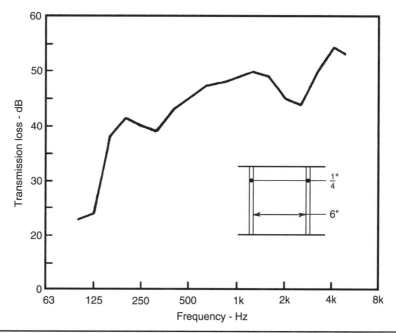

FIGURE 9-3 Transmission loss with heavy glass panes and large spacing between panes. (*Quirt, 1983*)

loss showed no such dip. Mass-air-mass resonant is largely a low-frequency effect; in fact, with certain glass panes at certain spacings, the resonant frequency is so low that it does not appear in the usual 63 Hz to 8 kHz measuring range.

The mass-air-mass resonant frequency can be estimated from the following equation:

$$f = 170\sqrt{\frac{m_1 + m_2}{m_1 m_2 d}} \tag{9-2}$$

where f = mass-air-mass resonant frequency, Hz
   m_1 = surface density of glass A, lb/ft^2
   m_2 = surface density of glass B, lb/ft^2
   d = spacing between glasses, in

The mass-air-mass resonant frequency has been calculated from Eq. 9-2 for many glass weights and spacings and plotted in Fig. 9-5. This figure shows that resonant frequencies are above 100 Hz only for lighter glass panes and smaller spacings; mass-air-mass resonance is primarily a low-frequency effect.

A prominent mass-air-mass resonance appears in the sound transmission-loss curve of Fig. 9-6 as measured by Quirt (Quirt, 1982). The notch at about 400 Hz is a good example of an acoustical hole in the glass. Near 400 Hz sound is attenuated about 5 dB less than at adjoining frequencies. Both the very small spacing of 1/4 in and the very light glass of 1/8-in thickness have purposely been selected to raise the mass-air-mass

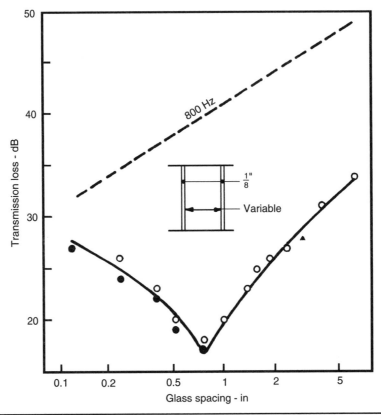

FIGURE 9-4   The mass-air-mass resonance effect.

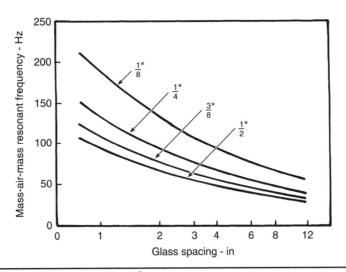

FIGURE 9-5   The mass-air-mass resonant frequency.

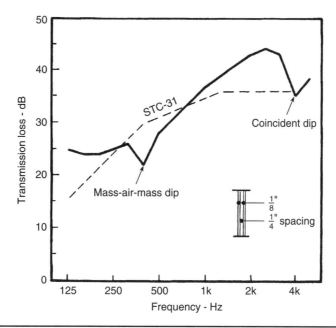

FIGURE 9-6 A comparison between mass-air-mass resonance and coincidence dip. (*Quirt, 1982*)

resonance to about 400 Hz to demonstrate its existence. For more practical double-pane windows, the resonance might be at too low a frequency to appear on a measured transmission-loss curve.

Acoustical Holes: Coincidence Resonance

The other dip at about 4 kHz in Fig. 9-6 is called a coincidence dip. It, too, can be classed as an acoustical hole, but is caused by an entirely different process. The flexural, bending vibrations of the glass panel interact with the impinging sound in such a way that an abnormal amount of sound is transmitted through the glass at the coincident frequency. When the phase of the pressure crests of the incident sound coincides with the vibrational crests of the panel, this coincidence effect results in lowering of the sound transmission loss. This means that sound at or near this frequency penetrates the window more easily. The frequency at which the coincidence effect occurs in a window may be estimated from:

$$f = 500/t \qquad (9\text{-}3)$$

where f = coincident frequency, Hz
    t = thickness of glass, in

The window of Fig. 9-6 employs glass 1/8-in thick. Equation 9-3 then becomes $f = (500)(8) = 4000$ Hz, which is close to the observed frequency of the coincident dip.

Coincident dips in windows can be minimized by using two glass panes. Each pane must have a different thickness; in this way, the resonant point of each pane occurs at a different frequency. When one pane is acoustically weak, the other pane is not. Clearly, a double-pane window in which both panes are of the same thickness would still encounter a coincidence dip problem.

Acoustical Holes: Standing Waves in the Cavity

The airspace gap between the two panes in a double-pane window is capable of supporting standing waves, much as in a room. Modes are associated with the length, height, and depth of the cavity as shown in Fig. 9-7. Axial modes strike two surfaces; tangential modes strike four surfaces; oblique modes strike all six surfaces. Because of the greater number of reflections encountered, the tangential modes and the oblique modes have lower energy levels (–3 dB and –6 dB respectively) below the axial modes (Morse and Bolt, 1944). The frequencies of all three modes may be calculated using Eq. 5-3.

Table 9-1 lists the frequencies associated with a window cavity having dimensions of 8 × 4 × 0.5 ft. The lowest axial-mode frequency associated with the 8-ft length, the (1, 0, 0) mode, is 70.6 Hz. The lowest axial-mode frequency associated with the 4-ft width of the cavity, the (0, 1, 0) mode, is 141.3 Hz. The lowest axial-mode frequency associated with the 6-in depth of the cavity, the (0, 0, 1) mode, is 1130 Hz.

It is important to know these frequencies in order to determine whether the absorbent material on the periphery of the space between the glass plates (the "reveals") is capable of damping these modal resonances sufficiently. If not damped, there is the possibility of minor acoustical holes appearing at certain frequencies. The thickness of this periphery absorbent is severely limited by the space available. The 70.6-Hz and 141.3-Hz axial modes are the lowest to be absorbed. The 1-in-thick glass fiber commonly used as cavity absorbent does not absorb these frequencies well, and there is no space for a Helmholtz

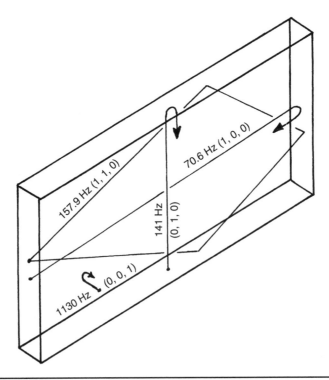

Figure 9-7 Resonances in the space between two glass panes.

p q r	Axial	Tangential	Oblique
1 0 0	70.6 Hz		
0 1 0	141.3		
1 1 0		157.9 Hz	
0 0 1	1130.0		
1 0 1		1132.0	
0 1 1		1138.0	
2 0 0	141.3		
2 0 1		1138.0	
1 1 1			1147.0 Hz
0 2 0	282.5		
2 1 0		199.8	
1 2 0		291.2	
0 2 1		1164.0	
0 1 2		2264.0	
2 1 1			1147.0
1 2 1			1147.0
2 2 0		315.8	
3 0 0	211.9		
0 0 2	2260.0		
3 1 0		254.6	
0 3 0	423.8		
2 2 1			1173.0
3 0 1		1149.0	

TABLE 9-1   Observation Window Cavity Resonant Frequencies
(Cavity Dimensions: 8 × 4 × 0.5 ft)

resonator absorber. Thus the cavity resonances at 70 Hz and 141 Hz will probably be significant in the finished window.

Effect of Glass Mass and Spacing

Figure 9-8 shows a compilation of many measurements of double-pane windows for 1/4-in and 1/8-in glass (Quirt, 1983). The figure shows the effect of glass mass, and spacing. The upper curve is for two 1/4-in glass plates, while the lower one is for two 1/8-in glass plates. The two curves indicate that there is a 3-dB gain in the STC rating for every doubling of the mass of the glass. The curves also show that doubling the spacing between panes results in approximately a 3-dB gain in the STC rating. Absorptive material should be placed in the interior areas between panes; this can add 2 to 5 dB of isolation. In the search for higher transmission loss with double-pane windows, several modest gains must be combined to get the best window performance.

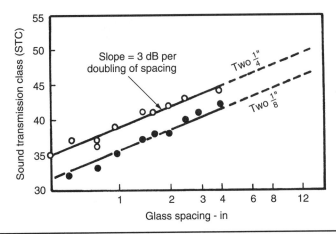

Figure 9-8   The effect of glass-pane spacing on STC rating. (*Quirt, 1983*)

Effect of Dissimilar Panes

Transmission loss can be improved by using two glass panes of different thicknesses. For example, consider two double-pane windows, both with approximately 4-in spacing, one with 1/8-in and 3/32-in glass and the other with 1/2-in and 1/4-in glass (as shown in Fig. 9-2). As both windows use dissimilar glass, specific evaluation of the dissimilarity factor is not possible. However, it is known that windows of different mass on the two sides will distribute the mass-air-mass resonances, resulting in a smoother transmission-loss curve.

The mass-air-mass resonance of the lighter window (the solid-line curve of Fig. 9-2) was found by calculation to be about 90 Hz, and that of the heavier window (the broken-line curve) to be about 57 Hz. These points are not discernible on the two curves. The coincidence effect involves only a single pane of glass. By calculation, the coincidence frequencies of the lighter window were found to be 4000 Hz for the 1/8-in glass and 5300 Hz for the 3/32-in glass. There is a pronounced dip in the solid-line curve in this region of the spectrum. The coincidence frequency for the 1/4-in glass was found to be 2000 Hz, and that of the 1/2-in glass was 1000 Hz. Small dips can be seen in the broken-line curve at both of these frequencies. Staggering these dips by using glass of different thickness tends to smooth the transmission-loss curve and minimize the effect of the acoustical hole. If both panes resonated at the same frequency, the dip would have been deeper. Therefore, insulation is improved when panes are of different thicknesses, and widely spaced. Windows with panes of the same thickness and with narrow spacing perform relatively poorly.

Effect of Laminated Glass

Glass may be laminated by placing a layer of polyvinyl butyral (PVB), often of 0.015-in thickness, between two layers of glass. Architectural laminated glass, the most familiar form, consists of two plies of glass bonded together by a plastic interlayer (usually PVB) under a pressure of about 250 psi and a temperature of 250° to 300° Fahrenheit. The laminating layer increases the weight of the glass, and modestly increases the transmission loss. This is shown in a comparison of equivalent unlaminated and laminated glass:

Unlaminated	Laminated
1/4-in glass: STC-29	Two plies 1/8-in glass: STC-33
1/2-in glass: STC-33	Two plies 1/4-in glass: STC-36

The use of laminated glass in double-pane windows results in greater sound transmission loss because of increased mass. There is also a small damping effect, which is an added advantage.

Effect of Plastic Panes

Plastic has the advantage of flexibility, and it does not shatter like glass. Such characteristics might suggest plastic instead of glass in an observation window in certain circumstances. The principal difference from the sound transmission-loss point of view is that the mass of plastic is about half that of glass and a corresponding double thicknesses of it would be required. When the masses are the same, plastic panes perform similarly to glass panes. Plastic sheets can be cold-bent to form convex windows. Light transparency of plastic is good, and optical distortion is minor. A convex form on the studio side would add a diffusing effect to incident sound and eliminate a potential "slap-back" echo.

Effect of Slanting the Glass

In many double-pane observation windows, one of the panes is slanted from vertical. As far as transmission loss of the window is concerned, slanting glass yields the same insulation performance as parallel glass if the average separation of the slanting glass is equal to the parallel glass separation. Based on a number of measurements, Quirt (1982) says "Nonparallel glazing does not appear to offer any significant benefits." On the other hand, from a room acoustics standpoint, it can be advantageous to slant one pane of glass so that reflections from the glass are angled downward to floor absorbers such as carpet. For example, this can control a potential reflection on the studio side of the window.

Effect of a Third Pane

Extensive measurements were made on windows with a third pane of glass (Quirt, 1983). Most of the three-pane results were very similar to those of the two-pane results. Small differences between three-pane windows and comparable two-pane windows showed a small advantage of the three-pane over the two-pane; this was attributed to a diminished coincidence effect. For most room window designs, the small advantage of a three-pane window over a two-pane window does not seem to justify the added cost and effort. In some cases, sealed triple-pane windows are used; for example, some airplanes use triple-pane windows.

Effect of Cavity Absorbent

The addition of a 1-in thickness of glass fiber around the perimeter of the interglass space was shown to be advantageous (Quirt, 1982). The improvement was limited to higher frequencies, as one would expect from the characteristics of the absorbent. These tests demonstrate the benefit of low-frequency absorbence in the cavity.

Thermal Glass

Two sheets of glass mounted together with an airspace of the order of 1/8 in to 1/4 in forms a very effective structure for reducing heat or cold transmission. The sound insulating properties of such glass are the same as those of a glass of the combined thickness. In other words, small airspaces have negligible acoustical effect, and such glass performs on the mass law alone.

Weak Windows (or Doors) in a Strong Wall

How much will a high-transmission-loss wall be compromised by mounting a window or door that has a lower transmission loss? Figure 9-9 illustrates a graphical method to make this determination. Notice that the figure uses a Transmission Loss (TL) rating instead of STC. STC is a single-number figure used to describe a series of TL measurements made at frequency intervals (usually 1/3 octave) throughout the frequency range. In other words, STC is obtained by a best fit of a standard STC contour to the actual TL measurements. The graph of Fig. 9-9 applies to a single TL measurement on the TL curve. This enables the building of an accurate point-by-point graph of the new TL curve, with the weak window/door mounted in the strong wall. This assumes that a measured transmission-loss curve is available for both the window/door and the wall.

The following approach uses STC figures on Fig. 9-9; this simplification limits accuracy but the results are still useful. For example, assume 1000 ft² of wall and 100 ft² of weaker window. Also assume that the wall area has an STC-50 rating, and the window an STC-30 rating.

1. Determine the ratio of the glass area to the wall area (100/1000 = 0.10), and find this value on the horizontal scale of the graph.

2. Determine the difference between the STC of the wall and that of the glass (50 dB – 30 dB = 20 dB); and find the 20-dB contour on the graph.

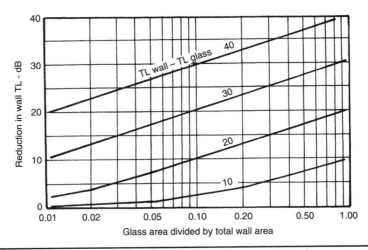

FIGURE 9-9 Graphical method for estimating the effect of a weak window in a strong wall.

3. The 0.10 vertical line intersects the 20-dB line; follow it to the left scale and read 10 dB.

4. Subtract this 10 dB from the wall STC (50 − 10 = 40).

The new STC of the weakened wall is thus STC-40, which we acknowledge is an approximate (but convenient) result.

Example of an Optimized Double-Pane Window

Figure 9-10 shows the measured transmission loss of a window that has STC-55 and a remarkably smooth curve (Cops, 1975). Surely the resonances of various kinds are well controlled to yield this smoothness. This window has dissimilar glass panes. One double-pane leaf consists of two 3/8-in glass panes with a 1/32-in PVB layer between them. The other pane is 5/16-in glass. The two panes are 12 in apart and an absorbent lines the periphery between the two panes. The important factors in this design are large separation, heavy but dissimilar panes, and absorbency. However, the absorbent produces a good improvement in sound insulation only when the mass of the panes is relatively light, which is the case for most windows. The lower curve in Fig. 9-10 is for a single 5/16-in

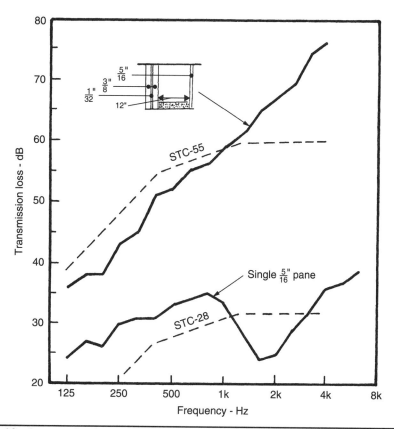

Figure 9-10 An example of an optimized double-pane window. (*Cops, 1973*)

glass pane to emphasize the great effect of adding the second, heavy double-pane to accomplish the STC-55 curve. A large coincidence dip greatly reduces the STC of the single pane.

Construction of an Observation Window

Observation windows separate two rooms where viewing is particularly important. For example, a window might be placed in the wall between a studio and a control room. When the window serves two acoustically sensitive rooms like these, the performance demands are extreme. Thus the detailed design and construction of the window is best left to experienced acousticians and carpenters. However, some elements of the design are common to all critical windows.

Assuming that a double wall or staggered-stud wall is used, the window frame is constructed in two halves. Each half is secured to one wall side with no rigid physical connection to each half; this preserves the wall decoupling. Foam strips or a nonhardening sealant plug the gaps between the window frames. The two glass panes use different thicknesses. The panes are mounted on resilient strips. Other details that are common to wall design, such as caulked seams, are employed.

Figure 9-11 illustrates the important features of a particular observation window design:

1. A strong frame of well-seasoned 2 × 10-in or 2 × 12-in wood is necessary for both frame and masonry walls. This frame must be made a part of the wall by packing glass fiber and acoustical sealant at the joints.

2. The glass panels must be fitted with foam or rubber strips obtained from the glazier. The strip bearing the weight of the glass along the bottom should deflect about 15% under load. The strips on the other three sides can be of lighter foam, as their only function is mechanical isolation and sealing.

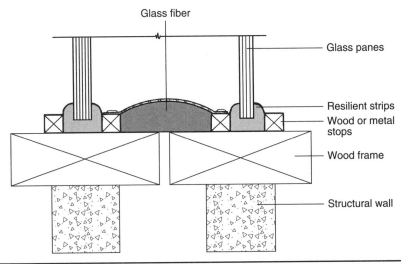

Figure 9-11 Construction details of an observation window.

3. The space between the inner stops should be filled with glass fiber and covered with black cloth or perforated material offering at least 15% in openings.

4. The outer stops should be screwed to the frame so that either glass can be removed for cleaning.

Proprietary Observation Windows

A well-constructed studio or other sound-sensitive space deserves the permanently precise treatment of openings that proprietary windows (and doors) provide. These openings are the worst place to economize, because deterioration with time is greatest in home-built windows and doors.

Figure 9-12 gives cross-sectional views of an STC-38 single-pane window and an STC-55 double-pane window manufactured by Overly. The transmission-loss curves for these two windows are shown in Fig. 9-13. Aside from a modest coincidence dip in the single-pane window, the curves show good performance.

Studio Doors

Common doors are probably the weakest link of all the sound barriers in studio construction. An acoustically weak door can defeat the isolation provided by an acoustically strong wall. As an extreme example, louvered doors provide little more isolation than an open door. Heavy doors provide better isolation than lightweight doors; for example,

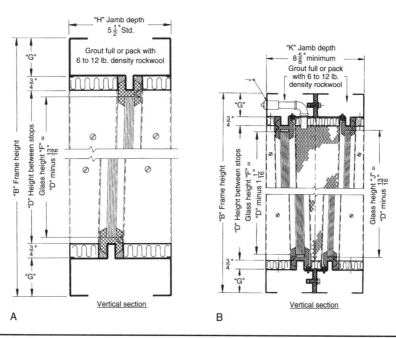

Figure 9-12 Overly proprietary windows. (A) STC-38 single-pane window. (B) STC-55 double-pane window. (*Overly*)

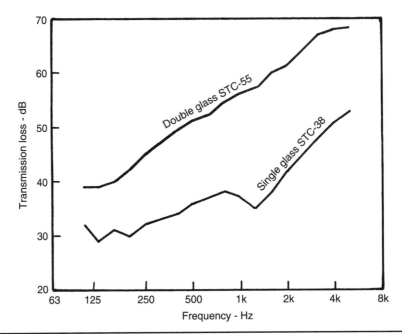

Figure 9-13 The sound transmission loss of two Overly windows (Fig. 9-12). (*Overly*)

a 1-3/4-in solid-core door is better than a same-thickness hollow-core door. Hollow-core household doors have STC ratings in the low 20s, and solid-core household doors have STC ratings in the upper 20s. Even these low values assume good weather stripping. Generally, metal doors perform better than wood doors. Manufactured acoustical doors and frames outperform other doors.

For any serious studio design, household doors are too inferior to be considered and must be replaced by heavier self-sealing doors. The hollow-core household door of Fig. 9-14 offers only about 15 dB of attenuation over much of the spectrum, but has an STC rating of 20. Solid-core doors are better with a rating of STC-27, but are still ineffective as a comprehensive barrier for noise. The sliding-glass door, STC-26, performs similarly to the solid-core door. The sliding-glass door is mentioned here because it is useful for drum booths and isolation booths. Although not recommended, a window can be placed in many door types without seriously compromising performance; for example, a 1/4-in plate glass window, 1-ft square and properly sealed, could be used.

The only truly satisfactory doors for sound-sensitive rooms are those metal doors and jambs designed specifically for the task and supplied by the industry. The Overly Model STC488861 door is shown in Fig. 9-15. It is 1-3/4-in thick with a surface density of 9.9 lb/ft^2 and rated as STC-48. It is an example of an engineered door that will yield superior results both immediately and over time.

The isolation of even a good door can be seriously compromised by air leaks. Therefore the perimeter of any door and frame must be gasketed and airtight. Common weather stripping can be used; several types are available including strips of felt or neoprene foam with metal backing. Some weather stripping uses magnets inside a plastic strip; the magnet presses the strip against a steel door, or a wood door with steel strips, to provide a tight acoustical seal. These types of seals work well on the tops and

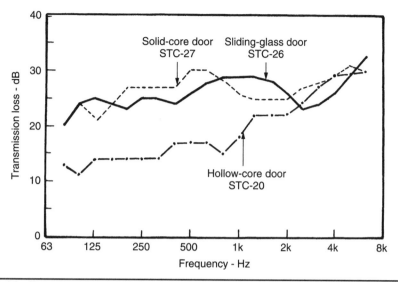

FIGURE 9-14 The transmission loss of three types of doors.

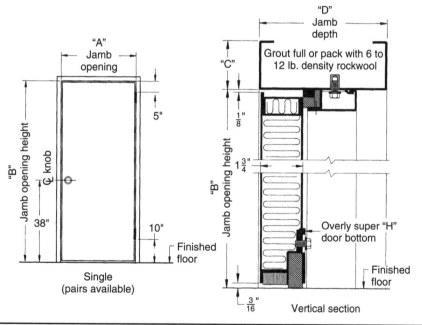

FIGURE 9-15 An Overly rated door. (*Overly*)

sides of doors; however, floor thresholds are subject to wear and tear from foot traffic, rolling dollies, and so on. Most kinds of weather stripping on thresholds can deteriorate rapidly.

For improved performance, a floor seal can use a metal strip that wipes a rubber seal across the threshold. The best door systems contain mechanisms that raise the internal threshold seal when the door is opened, and lower the seal tightly onto the floor thresh-old when the door is closed. Some doors use cam hinges; the door lifts up when opened, so the threshold seal does not wipe against the floor as it opens. This helps preserve the floor seal. When the door is closed, the cam helps press the door's weight against the surrounding seals. No matter which type of door seal is used, seals should be inspected periodically to make sure they remain tight.

Individual door design is important, but so is the way in which doors are placed in a room. For example, a door leading to a corridor should not face an opposite door on the corridor; this would allow sound to more easily pass between the rooms.

Sound-Lock Corridor

When the floor plan permits it, doors can be placed in sound locks; these short corridors allow the doors to act as independent sound barriers. Essentially, the two doors are placed in series, and can thus provide a much higher STC rating than a single door. Also, sound transmission is diminished because one door can be closed while the other is opened. Sound locks are a common feature in recording studios (see Fig. 12-1); this design shows a sound-lock corridor; the inner door leads to the high sound-level studio area and the outer door leads to the protected area.

A sound lock has two doors; if only one door is open at a time, the other closed door provides a degree of isolation. For optimal and long-lasting performance, the doors of the sound lock should be specially built acoustical doors. For improved performance, all inner surfaces of the sound-lock corridor should be very absorbent. In this way, even if both doors are simultaneously open, the absorption of the corridor will offer a degree of noise reduction. The ceiling could be suspended and use Tectum Ceiling Tile. The walls could use the same furred out (C-40 mounting) Tectum wall panel. Tectum is made of wood fibers and is quite resistant to abrasion. Alternatively, the upper parts of the walls could be constructed with glass-fiber batting behind stretched fabric or wire screen; sturdier wainscoting is used on the lower walls. As another option, Sonex absorbing panels could be used. The floor should be heavily padded and carpeted, but able to withstand the wear and tear of equipment dollies.

A short sound-lock corridor might not be sufficient. Anyone hurriedly exiting from either the control room or the studio would go quickly through two doors and a burst of sound could exit or enter. Moreover, a short corridor offers less absorption. Therefore, ideally, although it is never economical to do so, the sound-lock corridor should be as long as possible.

CHAPTER 10

Noise Control in HVAC Systems

Digital recording and reproducing techniques at both the professional and consumer levels demand low noise levels. In the acoustical design of most recording studios, control rooms, listening rooms, and other acoustically sensitive spaces, the primary background noise (provided the structure is isolated from outside noises) is the noise that is generated and distributed by the heating, ventilating, and air-conditioning (HVAC) system. These mechanical systems generally consist of an air-moving device with some form of attached ductwork. Each system has one major noise source (e.g., a fan) along with secondary sources (e.g., grille noise) throughout the system.

The control of HVAC noise can be expensive. When designing a new structure, the logical starting point is inserting in the air-conditioning contract a specification that a certain low noise level must be achieved. There must be knowledgeable liaison provided by the client to check the contractor's work during progress of the job. Someone on the client's staff should be designated for this liaison and should educate themselves in HVAC terminology and basic systems. An excellent way of doing this is to obtain a copy of the *American Society of Heating, Refrigerating, and Air-Conditioning Engineers (ASHRAE) Handbook,* and particularly study the portion on Systems and Fundamentals. No matter what approach is taken, controlling the background noise in sound-sensitive rooms often presents numerous acoustical challenges.

Selection of Noise Criteria

There are several different ways of stating HVAC noise design goals. The single number reading of a sound-level meter using the A-weighted scale is useful for noncritical systems. This dBA reading discriminates against the low frequencies, more or less the way the human ear does. This method is applicable only for undemanding situations. A weighting is discussed in Chap. 11.

Standard noise criteria (NC) or balanced noise criteria (NCB) contours are often used to establish the noise design goals for background noise in recording studios and other critically acoustically sensitive rooms. This method refers all measurements to the ultimate sensitivity of the human ear. The final question is whether the background noise can be heard in the room, or on recordings made in the room. Noise criteria contours are discussed in Chap. 11.

HVAC engineers have improved upon noise criteria by asking whether a noise sounds spectrally balanced. This introduces the issue of the sound quality of background noise. If a noise is audible in the studio or on a recording made in the studio, the quality of the noise becomes important. For example, a pronounced rumble or excessive hiss draws attention to the noise, while a carefully balanced rumble or hiss would be less noticeable. At least this much can be agreed upon: If the background noise is audible at all, a spectrally balanced noise is far preferable to a pronounced rumble or hiss.

HVAC engineers have devised their own NC/NCB criteria contours and refer to them as room criteria (RC) contours. The standard RC contours are shown in Fig. 10-1. The contours have been designed to evaluate audible artifacts such as rumbles (for example, caused by a fan) and hisses (for example, caused by air vents).

Fluctuations at air ventilation outlets are allowed (they are not allowed in the NCB standard). The use of RC contours is illustrated in Fig. 10-2. The fan noise spectrum peaks in the 63- to 250-Hz region. The air-vent noise is most prominent above 250 Hz. The attenuations of the two are separately adjustable, which is the key to balancing the sound. If the RC-35 contour is the design goal (as in Fig. 10-2), fan attenuation is adjusted to meet the RC-35 criterion, and the diffuser criterion can be independently adjusted to meet the RC-35 criterion. Of course, these adjustments are accomplished in the design stage.

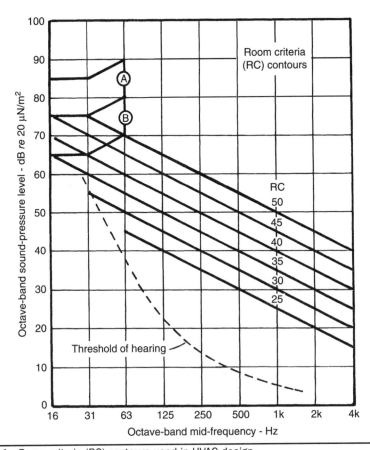

FIGURE 10-1   Room criteria (RC) contours used in HVAC design.

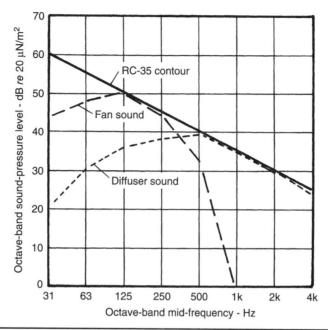

FIGURE 10-2 Example of balancing noise spectra with an RC contour.

In the NCB criteria contours of Chap. 11, the threshold of hearing is 10 dB or so below the NCB-15 contour. This means that with an NCB-15 studio criterion the background noise is audible, but barely. Sound-pressure levels of 70 or 80 dB magnitude below 63 Hz and which fall in areas A and B might induce audible rattles or noticeable vibrations in lightweight partitions and ceiling structures. A balanced background noise would surely be the economical approach to avoid over-designing fan or diffuser noise attenuation. The use of NCB and RC contours is described in *ANSI Standard 12.2 1995 (R1999), American National Standard Criteria for Evaluating Room Noise,* and in the *American Society of Heating, Refrigerating and Air-Conditioning Engineers (ASHRAE) Handbook.*

Fan Noise

The noise of a fan results from several noise-generating mechanisms. A siren-like tone results from interactions between the rotating and stationary members of the fan. The noise component consists of one or several pure tones. The blade-passage frequency of fan noise can be found from the product of the revolutions per second and the number of blades. Fans with less than 15 blades produce relatively pure tones that tend to dominate the spectrum. In addition to tone-like noise, fans generate random noise from the vortex of air created by fan motion. Noise is decreased when the inlet and outlet ducts are smooth so that air turbulence is reduced.

The specific sound-power level of pressure blowers and centrifugal fans is shown in Fig. 10-3. In general, pressure blowers are the noisier of the two. It is interesting to note that smaller pressure blowers are noisier than larger ones. Centrifugal fans are quieter than the pressure blowers, and larger ones are noisier than smaller ones.

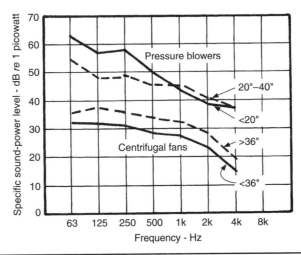

Figure 10-3 Sound-power levels of pressure blowers versus centrifugal fans.

HVAC Machinery Noise

Vibrations from HVAC or other mechanical equipment can produce low-frequency vibrations (less than 20 Hz) as well as higher-frequency noise. The former may only be felt, while the latter can be heard. Both must be reduced or eliminated in the design of an acoustically sensitive space. Vibrations can travel throughout a building's slabs and beams as structureborne noise and then reradiate as airborne noise elsewhere in the building. Clearly, because of the high noise levels found in most machinery rooms, great care must be taken when locating them in a building design; it is desirable to locate machinery some distance from the acoustically sensitive area. When the mechanical equipment room is in close proximity to acoustically sensitive rooms, comprehensive steps must be taken to ensure sufficient isolation.

A mechanical equipment room should be built with high sound isolation using concrete block walls, concrete slab floor, or similar materials. The interior walls and ceiling of a machinery room should be treated with thick absorptive material to lower ambient noise levels in the machinery room. Locating HVAC machinery on the roof of a frame structure or in the middle of a floor/ceiling span is not desirable because it will promote structural motion. Rather, machinery should be located near a structural wall or column for better support.

If HVAC machinery is mounted on a concrete slab, it is important to isolate the local HVAC slab from the structural slab. This is done during the pouring process by installing a compressed glass fiberboard or other isolation means between the two slabs. The vibrating machinery itself should be isolated from its own slab by placing it on resilient mounts such as steel coil springs, neoprene pads, or precompressed glass-fiber pads. Machinery should also be decoupled from any rigid structures such as pipes or ducts using nonrigid connectors. An example of the mounting of HVAC machinery in a mechanical equipment room is shown in Fig. 10-4.

In any machinery installation, it is important to specify vibration mounts of the correct design for the application. The vibration mount must be selected specifically for the

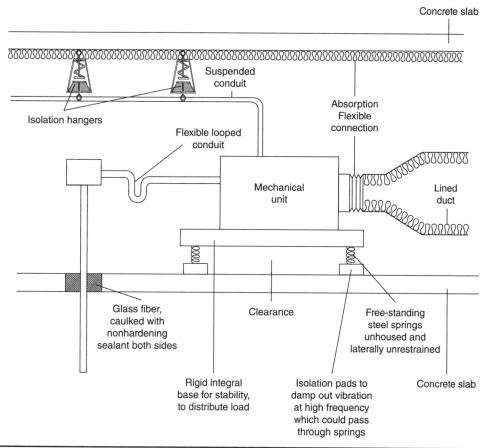

Figure 10-4 Example of the mounting of HVAC machinery in a mechanical equipment room.

characteristic of the machinery it will support. The mounting must be tuned to the appropriate frequency range; if the mounting is not correct, instead of decreasing the amplitude of vibrations, it can amplify them. Mounts should be selected so that the system's loaded natural resonant frequency is significantly lower (at least one-third) than the machinery's lowest driving frequency. This ensures a low transmissibility, that is, the ratio of the transmitted force to the machinery vibrating force. To achieve this, for example, pads and hangers must have sufficient static deflection when loaded. In most cases, the manufacturer of HVAC machinery will provide data to assist in correct mounting.

Plumbing Noise

Although a building's HVAC design often does not include plumbing, a few similarities are worth noting. Plumbing noise can be intrusive in an acoustically sensitive space. For that reason, water pipes should be routed away from these spaces, or in the least, because they can cause turbulent water noise, the number of fittings and valves should be minimized. Water pipes should be decoupled from the building structure with resilient mounts. Even when separated by distance, pipes can transmit noise to other rooms. Neoprene gaskets and sleeves and flexible connections can be used to reduce noise. In addition, water flow velocity should be kept low by using larger pipes; this reduces noise from turbulence. Pipes can be wrapped to reduce the radiation of noise. Special valves can be installed to eliminate water hammering noise, that is, the familiar sound created in a pipe when the flow of water is suddenly stopped.

Air Velocity and Aerodynamic Noise

The turbulent flow of air through HVAC ducts can generate significant levels of aerodynamic noise. Noise traveling through ducts can easily compromise the noise criteria specification of a room. Aerodynamic noise is most prominent from 200 Hz to 2000 Hz. Higher-frequency hiss noise (above 10,000 Hz) is created at vent outlets. Low-frequency noise (below 80 Hz) can be created by vibrating duct walls. Noise is higher at high air velocities. In particular, the noise level varies approximately as the 5th to 6th power of the velocity of the air flow. If air velocity is doubled, the noise level at the outlet will increase about 16 dB. Some authorities claim that in some cases air-flow noise varies as the 8th power of the velocity of the air flow. In this case, doubling air-flow velocity would increase the noise by 20 dB.

The quantity of air that must be delivered to a room is a basic design parameter that determines the size of the duct and air-flow velocity. Lower air velocities generate lower aerodynamic noise; low air velocity can be achieved by larger duct cross sections. For example, if the cross-sectional area of the duct is 1 ft^2, to deliver 500 ft^3/min requires an air velocity of 500 fpm (feet per minute). If the cross-sectional area of the duct is doubled to 2 ft^2, the air velocity is decreased to 250 fpm. Low-budget HVAC installations naturally incline toward smaller ducts, higher velocity air, and higher noise levels. An air velocity of 250 to 300 fpm is suggested as the maximum for sound-sensitive rooms.

Table 10-1 shows the relationship of noise criteria to air velocities at air supply registers and return grilles. Note that in a properly designed HVAC system, return velocities are always higher than supply velocities. The values in this table are valid for unobstructed lined ducts and registers with relatively large openings; these values are guidelines and may be replaced by more specific data supplied by the manufacturer.

Regardless of air velocities, smooth aerodynamics of the air-flow path results in lower noise. It is critical that the air distribution system present minimal resistance to air flow. Fittings that produce gusty and swirling air flow result in higher noise generation. Aerodynamic noise is produced at elbows, dampers, branch take-offs, sound

Noise Criteria	Air Velocity at Supply Register (fpm)	Air Velocity at Return Grille (fpm)
NC-15 to NC-20	250 to 300	300 to 360
NC-20 to NC-25	300 to 350	360 to 420
NC-25 to NC-30	350 to 425	420 to 510
NC-30 to NC-35	425 to 500	510 to 600
NC-35 to NC-40	500 to 575	600 to 690
NC-40 to NC-45	575 to 650	690 to 780

TABLE 10-1 Noise Criteria and Air Velocity Guidelines

traps, and the like. Small irregularities can cause significant increases in noise. For that reason, ducts should be designed with smooth and curved turns and branches, and transitions to different duct cross sections should be tapered and gradual. Some noise attenuators (described below) are designed with sharp bends and other features in a way that reduces overall sound energy, but can sometimes increase turbulence noise. It should be noted that noise travels with equal ease in both directions in ducts, regardless of air-flow direction. In other words, sound can easily travel upstream against the air flow.

HVAC Noise Attenuators

Seven types of HVAC noise attenuators are shown in Fig. 10-5. The lined duct (A) uses an absorptive lining to attenuate sound. Sound flows into a sound plenum (B)

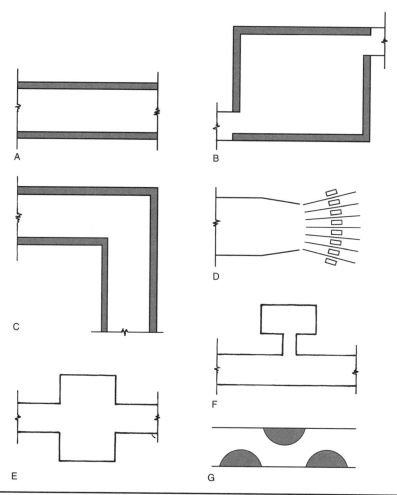

Figure 10-5 Seven types of HVAC attenuators. (A) Lined duct. (B) Lined plenum chamber. (C) Lined elbow. (D) Diffuser. (E) Reactive expansion chamber. (F) Tuned stub. (G) Silencer.

at the input. Some of the sound energy is absorbed by the lining, and what is left escapes through the outlet in attenuated form. The lined elbow (C) forces the sound to turn the corner through many reflections, each being attenuated by the absorptive lining. In fact, simply placing bends in a duct is among the most economical ways to reduce HVAC machinery and fan noise, and to reduce crosstalk between rooms. The diffuser (D) slows down the air velocity, which decreases noise level. The reactive expansion chamber (E) is an irregularity in the duct that reflects energy toward its source. The reverse phase of the reflection cancels some of the oncoming sound energy. The tuned stub (F) is designed to resonate at the frequency of a tonal component of sound in order to cancel some of its sound energy. It, too, uses a very sharp irregularity to reflect and cancel some of the undesired sound energy. A silencer (G) is a packaged attenuator that acts as a muffler; noise is attenuated as it flows around obstructions; absorbing material is usually protected by perforated metal sheets.

When an attenuating device is placed in a sound propagation path, the difference in sound levels before and after the device is referred to as insertion loss (IL). IL is thus used to rate the performance of noise attenuators. Aerodynamic noise created by some kinds of attenuators must be accounted for when considering their design and placement in the system.

Lined Duct

The noise energy of machinery can be attenuated by lining ducts with porous absorbing material. An internally lined duct is sometimes called a parallel baffle silencer because the absorbent is parallel to the air flow. This form of attenuator is widely used and easily implemented by the simple addition of absorbent to the duct system, particularly at turns. The performance of this type of attenuator depends on the length of absorbing treatment, thickness of absorbent, and the acoustical characteristics of the absorbent and its perforated-metal facing material. Attenuation is increased as the ratio of duct perimeter to cross-sectional area is increased.

The attenuation characteristics of three types of lined ducts are shown in Fig. 10-6; the same percentage of free area is maintained in all cases. The thinner absorbent in the closer spacings provides more high-frequency attenuation than the thicker absorbent in the wider spacings. In any of the three cases presented in the figure, the absorbent is too thin to realize much low-frequency attenuation. Similar information is presented in Fig. 10-7 for somewhat smaller ducts lined with 1 in of absorbing material, and with the attenuation expressed in dB/ft. Unlined metal ducts provide relatively little attenuation over distance. (Figures 10-6 through 10-14 are patterned after figures in the *ASHRAE Handbook*.)

For optimal isolation, air ducts should be decoupled from the building structure. For example, ducts can be mounted using suspension rods fitted with isolation hangers, and by placing resilient pads under supporting brackets. Internally lined ducts are highly efficient and cost-effective; however, linings can trap condensed moisture and this can promote growth of mold and bacteria. This in turn might cause sick-building syndrome or other air-quality health concerns. Care should be taken to limit moisture and to provide sufficient air flow to prevent this; these precautions are equally wise for unlined metal ducts.

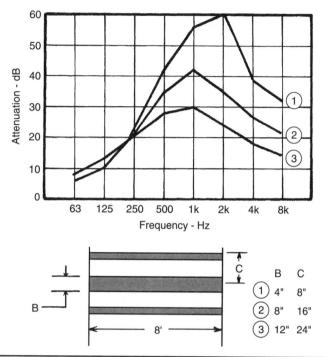

FIGURE 10-6 The attenuation characteristics of lined ducts.

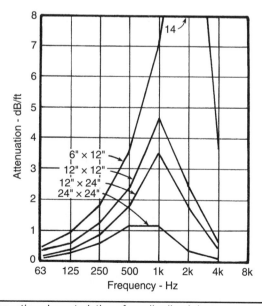

FIGURE 10-7 The attenuation characteristics of smaller lined ducts.

Lined Duct with Blocked Line-of-Sight

Figure 10-8 shows an adaptation of the straight-lined duct type of parallel baffle. By arranging a blocked line-of-sight, some of the economic advantages of thick panels are retained, as better high-frequency attenuation is achieved. If the noise is forced to go around the obstruction, it acts like a lined bend. These are proprietary attenuators and are available commercially.

Lined Duct and Length Effect

Figure 10-9 shows the effect of length on the attenuation performance of the lined duct arrangement of Fig. 10-6. It is interesting to note that the first 4 ft of treatment provides more attenuation than succeeding 4-ft sections. This is explained by the fact that some of the sound energy entering the duct is in cross-mode form, not parallel form. The cross-mode energy is rapidly absorbed in the first few feet of absorbent, leaving the parallel-mode energy to continue down the duct.

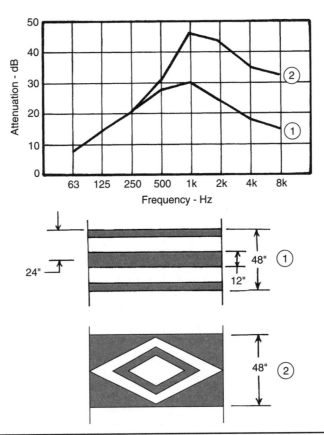

FIGURE 10-8 A comparison of blocked line-of-sight ducts and straight-through ducts.

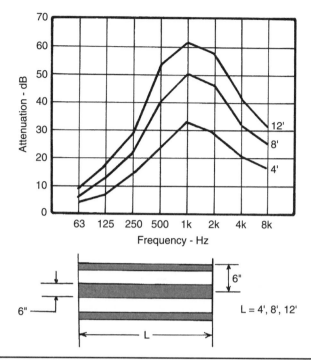

Figure 10-9 Attenuation characteristics of lined ducts.

Plenum Chambers

Lined ducts and other treatment downstream of machinery are effective, but are not always the most economical approach. A lined plenum of the form of Fig. 10-10 can be placed in the machinery room and will greatly attenuate the machinery noise before it reaches the duct system. A plenum chamber uses a relatively large interior volume; for example, its surface area should be at least ten-times greater than the cross-sectional area of the inlet. The greater the size of the plenum relative to the inlet and outlet, the greater the attenuation. Indeed, the cross-sectional area of the inlet and outlet should be as small as possible. With proper design, a plenum can provide attenuation over a wide frequency range.

A plenum provides attenuation as an expansion chamber in a reactive muffler and reflects energy back toward the source at both the entrance and exit irregularities. In addition, the plenum is lined with absorbing material which absorbs both sound energy coming in and sound energy reflected from the exit irregularity. Attenuation of the plenum can be increased by increasing the ratio of the cross-sectional area of the plenum to the cross-sectional area of the entrance and exit openings. Thicker absorbent also increases the attenuation. Often a room near the HVAC machinery is designed to operate as a plenum. In some designs, a corridor can be used as a plenum.

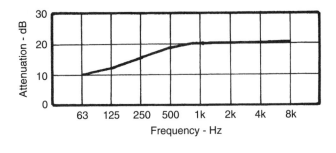

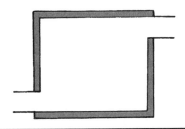

FIGURE **10-10** Attenuation characteristic of a plenum chamber.

Lined Elbows

One way that sound travels around a duct elbow is by the reflection of sound waves from the bend surface. This effect is most important at frequencies at which the wavelength is small compared to the duct dimensions. Reflected sound is absorbed if the bend is lined with absorbing material. The general attenuation characteristics of the lined bend are shown in Fig. 10-11. Thicker lining will increase this attenuation. Turning vanes can be placed in elbows such as these to decrease aerodynamic noise.

Diffusers

The sound attenuation of a typical diffuser (vent opening) is shown in Fig. 10-12. Attenuation is achieved in the process of reducing the velocity of an air stream. This is important because the noise of an air stream is proportional to the stream velocity. A small reduction in the velocity results in a large reduction in air-stream noise. This is the only device considered so far that is more effective at reducing noise at low frequencies than at higher frequencies.

Reactive Expansion Chamber

Any irregularity in the characteristics of an electrical transmission line reflects energy back toward the source. A duct carrying sound acts in the same manner. The expansion chamber of Fig. 10-13 presents an irregularity in the acoustical characteristics of the duct. In fact, both its input and the output reflect energy back toward the source. One object is to make the expansion chamber width a quarter wavelength at the frequency

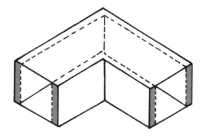

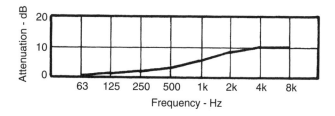

FIGURE 10-11   Attenuation characteristic of a lined elbow.

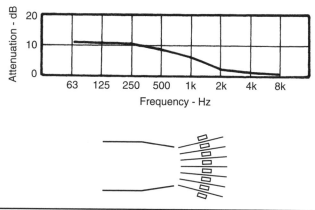

FIGURE 10-12   Attenuation characteristic of a diffuser.

of interest so that reflections from the output will cancel energy because of its opposite phase.

An expansion chamber having the proper dimensions to cancel energy at one frequency will have no such effect on other frequencies. Therefore, it is necessary to tune the chamber to attenuate single-frequency components of the noise. This tuning will place the first maximum of attenuation at the quarter-wavelength point. As a dividend, attenuation peaks through the spectrum will also appear at frequencies of odd multiples of quarters of a wavelength. Several expansion chambers might be used in series, each tuned to a different noise component. No acoustical material is needed to obtain this attenuation.

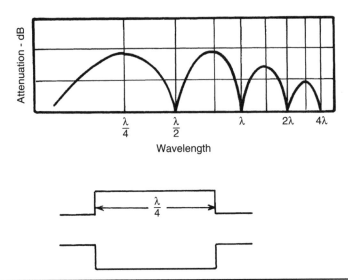

FIGURE 10-13   Attenuation characteristic of a reactive expansion chamber.

Tuned Stub

Another type of reactive attenuator relies on an acoustically tunedstub that opens into the duct, as shown in Fig. 10-14. This resonator attenuates only a narrow band of frequencies, and has no multiples appearing in the spectrum. Even a small unit of this type can produce 40 or 50 dB of attenuation at the frequency of resonance with negligible effect at other frequencies. It offers little obstruction to air flow. Several stubs can be placed in series to attenuate other noise components.

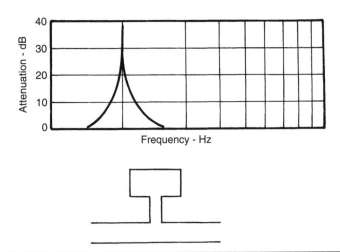

FIGURE 10-14   Attenuation characteristic of a tuned stub.

Silencers

Silencers, also called mufflers, are designed to provide high attenuation over a wide frequency range, using a relatively small unit. Total absorption increases as the length of the silencer increases. Silencers use a series of internal baffles with a blocked line of sight and are treated with absorptive material; because of their design, sound energy is exposed to a large surface area of absorption. In some critical acoustical designs, silencers cannot be used because of the aerodynamic noise they can introduce. This effect can be reduced by placing silencers as far away from acoustically sensitive rooms as practical.

Active Noise Cancellation

With active noise cancellation, a loudspeaker is driven by an out-of-phase signal to cancel noise using destructive interference. The technique is shown in Fig. 10-15. Noise traveling down the duct is sensed by a microphone and processed, a new signal (including a reversal of phase) is generated, and that signal is amplified and used to drive a loudspeaker downstream. The signal from the loudspeaker theoretically cancels the noise, thus achieving attenuation. An error microphone further downstream can be used to sense remaining noise, and used in a feedback control system to adaptively optimize the cancellation. Such systems have been successfully applied to single points. For example, active noise cancellation can be used in an air duct to decrease low-frequency noise from an HVAC fan. It may be possible to attenuate ventilation noise or tones by 30 to 40 dB. As another example, the noise level at a factory workstation might be so high that conversation is difficult; a usable reduction can be obtained at the workstation with active noise cancellation. Sound-canceling headphones have also been quite successful. The principle of active noise cancellation is being applied to HVAC systems on a limited basis.

Natural System Attenuation

There are some energy losses throughout an HVAC system even without duct lining or sound traps. This amounts to a natural attenuation, which should be accounted for, lest the system be overdesigned. Not all of the acoustical energy generated by a fan, duct fittings, and other components actually reaches a given room. Energy division at branch take-offs, the vibration of duct walls, and reflection losses at bends and duct outlets all contribute some attenuation. For example, a round duct with or without thermal insulation has a natural attenuation of about 0.03 dB/ft at 1 kHz, rising to about 0.1 dB/ft at

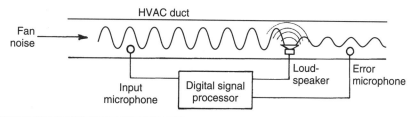

FIGURE 10-15   Active noise cancellation.

higher frequencies. These types of natural attenuation should be taken into consideration when specifying a system.

Ductwork Design

When designing HVAC systems, practical experience and attention to detail are vital. For example, a person unfamiliar with acoustical noise problems might overlook the fact that sound travels bidirectionally through ducts, and thus might place two air outlets as in Fig. 10-16 (A). The resulting crosstalk between the rooms would seriously degrade the insulating effectiveness of a wall, for example, between a control room and a recording studio. Similarly, any short and direct ducts between rooms would introduce a crosstalk path. Widely separating the outlets as in Fig. 10-16 (B) would help maintain isolation. The isolation between outlets is greater in (B) than in (A), and if the duct is lined, the attenuation might be sufficient. In critical designs, the ductwork between adjacent rooms must be independent, as in Fig. 10-16 (C). Whenever possible,

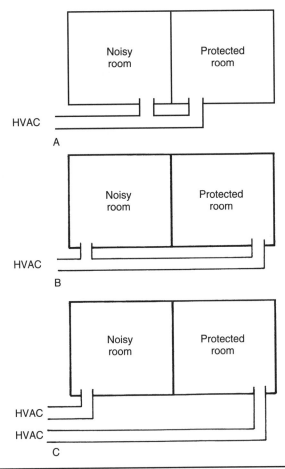

FIGURE 10-16 Room duct routing. (A) Poor routing. (B) Acceptable routing. (C) Improved routing.

ductwork should be designed with lined ducts, elbows, inline silencers, and extended duct paths between rooms.

High-performance duct designs incorporate many special features. For example, when ductwork passes through a noisy room, even without a vent opening, noise can easily enter the duct through the duct wall and travel to another noise-sensitive room. For that reason, the duct length in the noisy room should be isolated, for example, enclosed in gypsum-board drywall or plywood. As another design example, in some cases it may be advantageous to remove grille or vent covers; this may reduce aerodynamic turbulence noise, and also increase duct-end reflection, which will increase low-frequency noise attenuation.

CHAPTER 11

Room Performance and Evaluation

A s with any endeavor, we need a way to asses the acoustical performance of a room. Acoustical design is not an exact science; rather, it is a science and an art. Even the most careful designer cannot guarantee that the finished room will precisely match the design goals. While much mathematical modeling can be done in the design phase, a project is not complete until the result has been comprehensively measured and evaluated.

Both objective and subjective methods can be used to measure and evaluate acoustical properties. On one hand, some aspects of acoustical performance can be objectively quantified and measured with relative ease. For example, instruments such as sound-level meters can determine the level of noise in a room. These measurements can be organized by a standard measuring protocol, for example, to rate the room's background noise.

On the other hand, other aspects of acoustical performance are harder to quantify. For example, a concert hall must impart a sense of spaciousness. While certain variables relating to spaciousness can be measured, it is often better to listen to music performed in a hall and subjectively rate its spaciousness. Likewise other criteria can be subjectively evaluated to determine the hall's sound quality. As with objective evaluations, subjective evaluations can be made consistent by using a standard set of evaluation criteria, and expert listeners.

Sound-Level Meters

The sound-level meter, an important tool for a site noise survey, is a simple instrument in principle. It incorporates a microphone to convert the sound pressure changes to voltage changes, an amplifier to amplify the voltage changes, a frequency-weighting network, filters to reject signals at frequencies outside a selected band, a calibrated attenuator to extend the range, and an indicating meter or read-out. Obviously, the quality of the instrument depends on the quality of the integral parts and the stability of calibration.

The sound-level meter accepts the noise disturbances in the air and reads out the sound-pressure level (SPL). The sound-pressure level is the sound pressure expressed in decibel above the standard sound pressure of 20 micropascals, which is close to the threshold of hearing of the human ear. This is a physical measurement and is not a direct measure of loudness, which is a psychophysical parameter. However, loudness can be

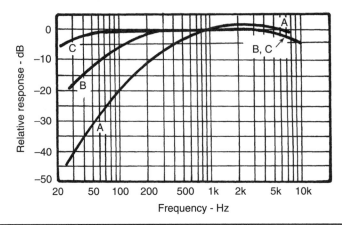

Figure 11-1 Weighting curves for sound-level meters.

calculated from sound level measurements using other concepts in the process. Sound-pressure level is of great value in manipulating the tangibles and intangibles of noise.

Most sound-level meters incorporate three standard weighting networks, which change the frequency response as shown in Fig. 11-1. Their purpose is to generally correlate the readings with human perception of loudness. The A-weighting network is approximately the inverse of the 40-phon equal-loudness contour of the human ear. To approximate the ear's insensitivity to low-frequency signals especially at low listening levels, the A-weighting network rejects low frequencies. The B- and C-weighting networks are approximately the inverse of the 70- and 100-phon equal-loudness contours respectively. B weighting moderately rejects low frequencies while C weighting only slightly rejects low frequencies. It follows that the A-weighting network is intended for use in measuring low sound levels, B weighting for intermediate sound levels, and C weighting for high sound levels. These will make readings conform somewhat better to human judgments of loudness but as noted, sound-level meters do not measure loudness. Z (zero) weighting is used to describe a flat frequency response from 10 Hz to 20 kHz ± 1.5 dB. Z weighting is specified in the IEC 61672 standard, introduced in 2003. It replaces previously used proprietary "flat" networks that used arbitrarily placed low and high cutoff frequencies or did not define the linear range of the device.

Readings taken with a weighting network are thusly designated; for example, if an A-weighting network is used, the measurement is expressed in dBA. When used with weighting filters, an SPL meter cannot analyze specific frequency content. For example, a rumbling noise and a hissing noise may both have the same dBA reading, but sound very different. However, an estimate of the noise response can be obtained. For example, if a dBA reading is lower than a dBC reading, the input signal has significant low-frequency content. Conversely, dBA and dBC measurements that are equal to within 3 dB would indicate higher frequency content in the measured sound. Any dBA-weighted measurement cannot exceed a dBC-weighted measurement by more than 3 dBA; if so, there is an error in the measurements. Table 11-1 shows the standard response corrections employed in the A- and C-weighting networks. None of the weighting networks apply filtering at 1000 Hz, so all weighted and unweighted measurements at 1000 Hz should be equal.

Octave Band Center Frequency (Hz)	A-Weighting Correction (dB)	C-Weighting Correction (dB)
63	−26.2	−0.8
125	−16.1	−0.2
250	−8.6	0
500	−3.2	0
1000	0	0
2000	+1.2	−0.2
4000	+1.0	−0.8
8000	−1.1	−3.0

TABLE 11-1   A- and C-Weighting Corrections

Sound-level readings, often expressed in dBA, are commonly used to estimate environmental noise levels from sources such as vehicle noise, and are used for ordinance code enforcement. Table 11-2 shows examples of environmental sound-pressure levels measured in dBA. Sound-level readings in dBC are more often used to measure mechanical noises. The B-weighting curve is rarely used today. Because Z weighting defines a flat frequency response, this measurement yields the overall decibel level.

As noted, single broadband SPL readings do not provide detailed information on the spectral response of the noise. Thus sound-level meters often contain narrowband filters; these filters reject frequencies outside the selected audio band. This allows frequency analysis of input signals; this is useful because many acoustical characteristics

Sound Source	SPL (dBA)
Air raid siren (threshold of pain)	120
Rock concert in audience	110
Subway train passing by platform	100
Truck passing by sidewalk	90
Heavy traffic passing by sidewalk	80
Medium traffic passing by sidewalk	70
Busy urban office	60
Suburban area in day	50
Suburban area at night	40
Rural area at night	30
Broadcast studio	20
Audiometric booth	10
Threshold of hearing	0

(from Cowan, 1994)

TABLE 11-2   Common Environmental Noise Levels

are frequency dependent. For example, a partition may be effective at blocking high frequencies, but perform poorly at low frequencies. A broadband SPL reading may not show, for example, that a noise had considerable low-frequency content. A more detailed frequency analysis would reveal that.

Many sound-level meters thus have filters with octave, 1/3-octave, or narrower width. The center frequencies of these bands are logarithmically spaced inside each band; each successive center frequency is twice that of the preceding center frequency. The most commonly used center frequencies for octave bands are: 63, 125, 250, 500, 1000, 2000, 4000, and 8000 Hz. Sound-level meters also allow different measuring speeds with different integration times, for example, slow (1 sec/reading), fast (125 msec/reading), and impulsive (35 msec/reading). The slow speed is typically used for environmental measurements and the faster speeds for measuring discrete events. The American National Standards Institute (ANSI) publishes standards on sound-level meters and sound measurements; for example, *ANSI Standard S1.43-1997 (R2007)* describes specifications for sound-lever meters, and *ANSI/ASA Standard S1.6-1984 (R2011)* defines frequencies, frequency levels, and band numbers for acoustical measurements.

Brüel & Kjaer manufactures several sophisticated sound-level instruments. For example, Fig. 11-2 shows the Brüel & Kjaer Type 2236 sound-level meter. The Type 2236 is a precision integrating instrument that can display both peak and root-mean-square (rms) values simultaneously because it has two independent channels. The meter is operated and results are displayed via an LCD screen. An interactive menu guides the user through the measurements and warns against changing the set-up parameters once a measurement has started. Three user-definable LN values are available, and cumulative distributions of the results provide basic statistics. A real-time clock marks the date and time of every measurement. There is an automatic start feature that allows automatic starting of a run. L_{eq}, L_{max}, L_{10}, and L_{90} can be stored as a set, and 21,600 sets can be stored in memory. These descriptors are discussed later. This is equivalent to six hours of logging at one-second intervals. The striking shape of the body of the instrument, with the microphone at the tip, essentially eliminates distortions of the sound field due to physical interference of the instrument case.

Quest Technologies and Larson-Davis also manufacture a variety of sound-level meters. These manual or automatic, integrating, and data-logging instruments offer different weighting sound levels and one- and 1/3-octave bands. It is possible to create and store multiple files without the necessity of downloading. Sound-level meters such as these are capable of handling almost any industrial, environmental, or community noise measurement.

Sound-level meters of great sophistication are not always essential for a noise survey of a site, but they make it possible to perform a fast, complete, and accurate job. Measurements are fast and simple so they encourage taking measurements at more environmental sites, or for extended hours or days for better coverage. A more complete assessment of a site's noise exposure can only make the design of a sound-sensitive room that much better.

Spectrum Analyzers

As noted, many sound-level meters contain narrowband filters to allow frequency analysis of input signals. However, spectrum analyzers allow a more thorough frequency

FIGURE 11-2   A Brüel and Kjaer Type 2236 sound-level meter.

analysis. For example, they use a set of ten octave bandpass filters to allow examination of sound-pressure levels for specific frequency regions. Analyzers also provide bands such as 1/3-octave, 1/10-octave, or more narrow; in many cases, 1-Hz precision is provided. As with SPL meters, octave bands are set so that each upper-frequency limit is twice that of the lower-frequency limit. Thus, each successively higher band is twice the width of the previous band; adjacent bands overlap at −3-dB points. The center frequency is the geometric mean. Spectrum analyzers are available in a variety of handheld and computer-based models. When implemented on laptops or tablets, data from spectrum analysis software can be joined with other software to provide powerful measurement and diagnostic tools.

The Noise Survey

There are many financial aspects involved in choosing a site for a building such as land value, building cost, and space rental. Clearly, more desirable or profitable properties will demand higher prices. It is also expensive to keep outside noises from getting into the building and vice versa. To minimize construction or remodeling costs, it is important to compare potential locations on the basis of a 24-hour noise-level survey. A noise-level survey is a simple procedure, but one that yields extremely valuable information.

As a learning experience, nothing compares to taking a noise survey "by hand" with a sound-level meter that provides only sound levels. This might seem elementary and time-wasting, but it will give those running the survey a fundamental understanding of the process. Moreover, although it is laborious, this kind of survey is inexpensive.

First, a single point outdoors is selected that is close to the proposed studio site, and which seems to sample the known noisemakers of the neighborhood. A 24-hour period of measurements is then scheduled for this point. A sampling period could be hourly, or every 15 seconds. It is suggested that a measurement be taken every minute on the minute. This yields a set of 1440 samples, which should give a reasonable picture of the environmental noise of the neighborhood. Readings should be taken with the A-weighting network in the circuit. As noted, these are called dBA measurements.

A suggested data sheet template for recording noise measurements is shown in Fig. 11-3. Sound-level readings will not be written down as numbers, but only as slashes in the appropriate column in which the levels fall, marked in the lower part of

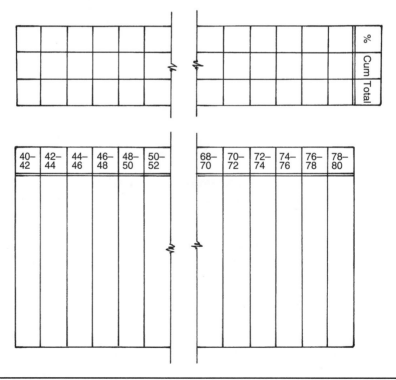

FIGURE 11-3   A template for collecting noise survey field data.

the template. The fifth slash in any given column should be a diagonal to help in the counting later. Each column embraces 3-dBA readings, such as 48–50, 72–74, and so on. Little final accuracy will be lost through this simplified 3-dBA coarseness. Every minute a slash will be added to this form in the appropriate column. This is the only duty until the 24-hour measurement period is completed.

At the end of the 24-hour period, there is sufficient data on hand to build a statistical distribution curve for that locality. The upper part of the template in Fig. 11-3 can now be completed. The Total column is filled in, with the total number of slash marks in each column. When the Total column is full, entries are made in the Cumulative column. The total number of slashes in the 78–80 column is added to the total of the 76–78 column, and recorded in the associated box in the Cumulative column. The total of the 74–76 column is then added to the sum of the 76–78 and the 78–80 totals, and so forth. The last Cumulative entry will be the total of all slash marks on the page.

Next, the Percent (%) column is completed. The cumulative figure in the first box is divided by the total number of slashes, divided by 100 to express the figure as a percentage, and the result is recorded in the first Percent box. The second cumulative amount is then divided by the total in the Cumulative column, expressed as percentage, and recorded in the second box in the Percent column, and so forth.

Everything is now available to draw a distribution curve such as the example shown in Fig. 11-4. The first point on the distribution curve is found by plotting the first

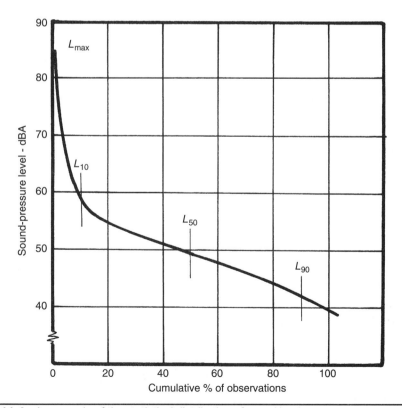

FIGURE 11-4 An example of the statistical distribution of sound-level measurements.

percent value in the percent box against the 79-dBA level. The second plotted point is the second percent value against 77 dBA, and so forth. This is a significant curve that gives a complete statistical picture of the environmental noise at that location, for that 24-hour period in the week.

Environmental Noise Assessments

A variety of descriptors have been devised to measure and evaluate environmental noise. These include L_{eq}, L_{dn}, L_{max}, L_n, and SEL; they are measured in dBA. L_{eq} is defined as the continuous equivalent sound level; it is a sound-pressure level that would contain the same energy as the fluctuating sound (such as traffic levels) that is being monitored over the evaluation period. The evaluation time period is indicated in parenthesis; for example, $L_{eq(1)}$ refers to a 1-hour period and $L_{eq(24)}$ refers to a 24-hour period. L_{dn} (sometimes known as DNL) is a day-night equivalent sound level; it is a continuous $L_{eq(24)}$ measurement with 10 dBA added to levels recorded from 10 PM to 7 AM. It is intended to assess noise level, for example, near airports; the 10 dBA value is added to account for the fact that noise is more intrusive during sleeping hours. Some acousticians argue that L_{dn} does not fully account for isolated loud events such as aircraft flyovers.

When recording data points in a template (such as shown in Fig. 11-3), there will be times when airplanes fly overhead, trains pass, and sirens wail, giving levels far above the usual background values. The maximum values must be recorded elsewhere, as the noise assessment at this point is not complete without them. From all these extra high values, one is the highest and it becomes L_{max} for the survey. It is possible that this is the most important reading taken because noise such as this could penetrate the walls of a studio, albeit infrequently. All of the maximum values together can be analyzed as a separate problem. If a train passes several times during a 24-hour period, perhaps this is the one noise that dominates decisions of wall construction. On the other hand, if a plane passing overhead is the only one heard for a week, it should have a lesser influence on the wall construction.

Noise levels over time at a site can be quantified using the L_n designation. L_n defines a percentile sound level where n (0 to 100) is the percentage of the measurement time period when a certain sound level measured in dBA is exceeded. The descriptors L_{10}, L_{50}, and L_{90} are often used to indicate noise levels with respect to thresholds. For example, $L_{10} = 70$ dBA describes an SPL measurement where a value of 70 dBA was exceeded 10% of the time. L_{10} indicates the loudest noises over time. The relationship of L_{10}, L_{50}, and L_{90} is shown in Fig. 11-5. In addition, the measuring time period is quoted; for example, $L_{10(1)}$ denotes a 1-hour time period. L_n provides a way to document fluctuations in sound-level measurements; for example, a large difference between L_{10} and L_{90} readings over a 1-hour period would indicate significant changes in a noise environment.

The statistical distribution of sound levels can be easily plotted. A distribution graph (such as shown in Fig. 11-4) is indefinite on the L_{max} value, but is very definite on L_{10} (the noise level exceeded 10% of the time), L_{50} (the noise level exceeded 50% of the time, that is, the median sound level), and L_{90} (the noise level exceeded 90% of the time). These data are very important in deciding how much isolation studio walls must provide.

The question will arise, how will L_{10}, L_{50}, and L_{90} vary throughout the day? This is where integrating instruments excel. Figure 11-6 illustrates one way of presenting such information. Note that with these SPL measurements, the frequency response of the noise is not taken into account.

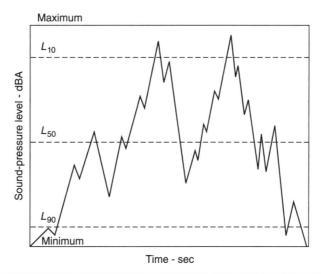

FIGURE 11-5 The descriptors L_{10}, L_{50}, and L_{90} used for noise assessment with level thresholds.

The Sound Exposure Level (SEL) defines and measures the total energy of a single noise event such as a train passing or an aircraft flyover. For easier comparison, the SEL compresses noise events into a 1-second period, even though the actual event time may be greater than 1 second. Because of this compression, SEL values lasting more than 1 second are higher than other measurement values. The Noise Exposure Forecast (NEF) can be used to estimate single noise events.

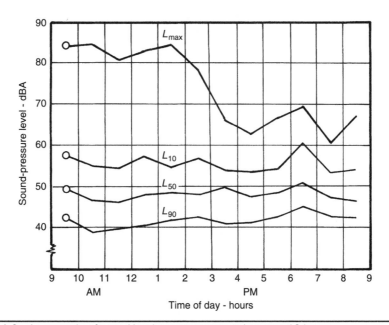

FIGURE 11-6 An example of sound-level measurements taken over 12 hours.

Measurement and Testing Standards

A variety of standards have been developed for assessing acoustical noise environments. The American National Standards Institute (ANSI) and the American Society for Testing and Materials (ASTM) publish documents that define and describe these standard methods. When these standards are followed, the results of noise measurements can be considered to be credible. Some commonly used standards are: *ANSI S1.13-1971 (R 1986) American National Standard Methods for the Measurement of Sound Pressure Levels*; *ANSI S12.9-1988 (R 2003) American National Standard Quantities and Procedures for Description and Measurement of Environmental Sound, Part 1*; *ANSI S1.26-1978 (R 1989) American National Standard Method for the Calculation of the Absorption of Sound by the Atmosphere*; and *ASTM E1014-84(1990) Standard Guide for Measurement of Outdoor A-Weighted Sound Levels*.

ANSI S1.13-1971 (R 1986) American National Standard Methods for the Measurement of Sound Pressure Levels defines survey, field, and laboratory methods to measure sound levels. The survey method uses a handheld SPL meter to measure sound levels in the environment. To obtain source-only readings, level readings are adjusted when large differences (4 to 15 dB) are noted when a sound source is on or off; this is described in more detail below. The field method uses an octave or narrowband analyzer to perform spectrum analysis of the noise level; the measuring microphone must be mounted on a tripod or suspended, and observers should be a distance away. The laboratory method is used in controlled environments such as anechoic or reverberation chambers. The standard also describes fluctuating or impulsive sounds, distances from obstacles, sizes of obstacles, room interior measuring placement, atmospheric and terrain conditions, and other factors that influence measurements.

ASTM E1014-84(1990) Standard Guide for Measurement of Outdoor A-Weighted Sound Levels describes simple guidelines for measuring outdoor noise levels. An SPL meter is used; a headphone output is recommended so the operator can detect unwanted sounds such as wind noise. Measurements should be taken during noisy and quiet periods of the day, when a sound source is operating and when it is not. The number of instantaneous measurements taken must be at least 10 times the range of the reading levels. For example, if the range is 5 dB, 50 measurements must be taken.

Recommended Practices

When taking noise measurements, a number of recommended practices should be followed. For example, a calibrator should be used to check the accuracy of any sound-level meter. A calibrator fits over the microphone and generates a tone (for example, a 1-kHz sine wave) or a series of tones at a certain sound-pressure level. A meter should accurately read the SPL of the tone emitted by the calibrator, or be adjusted to an accurate reading. A meter should be checked before it is used, and afterward as well to ensure that its calibration has not drifted. Adjustments to calibration may be necessary if a weighting network filter is used. When a 1-kHz calibration tone is used, no adjustment is needed because both A- and C-weighting networks are equal (0 dB) at that frequency.

For outdoor measurements, a properly fitting wind screen should be used to reduce wind noise at a measuring microphone. Even then, measurements should not be taken when wind speeds are greater than 15 mph. Measurements should not be taken while it is raining or even if the ground is wet after a rain; the added sound of rain or wet surfaces may affect the readings, for example, traffic sounds are different on wet pavement.

Difference in dBA between SPL Measured with Sound Source Operating, and Background SPL	Correction in dBA to be Subtracted from SPL Measured with Sound Source Operating to Obtain SPL from Source Alone
4	2.2
5	1.7
6	1.3
7	1.0
8	0.8
9	0.6
10	0.4
11	0.3
12	0.3
13	0.2
14	0.2
15	0.1

[from ANSI S1.13-1971(R 1986)]

TABLE 11-3 Standard Corrections for Measured Ambient Sound-Pressure Levels

Other meteorological conditions such as relative humidity, temperature, and barometric pressure can affect sound-level readings. They should be noted in the report, and corrections applied as necessary. The measuring microphone should be placed at least 3 to 4 ft away from any reflecting surface including the ground, as well as the observer, to avoid reflections; sound-pressure level increases significantly (up to 6 dB) near a reflective surface.

When measuring the noise level of a sound source, if the SPL with the source operating is between 4 and 15 dBA of the level when the source is not operating (background noise level), the measured level must be adjusted. In particular, the background noise level is significant and must be subtracted from the combined noise level to accurately determine the noise level of the source alone. The amount subtracted depends on the difference between the two measured levels (source on or off). The corrections are listed in Table 11-3. If the difference is 3 dB or less, one can conclude that the sound level of the source is equal to or less than the background sound level; therefore, the sound level of the source cannot be accurately determined from the reading. A final report should include all pertinent information, including a sketch of the site showing measured locations and distances of sound sources and measurement positions.

Noise Criteria Contours

The standard way to define the noise level within any structure is to use balanced noise criteria (NCB) contours; these are shown in Fig. 11-7. These criteria curves plot octave SPL with respect to frequency and show different levels of maximum permissible noise. The curves slope downward with frequency because the ear is less sensitive to low

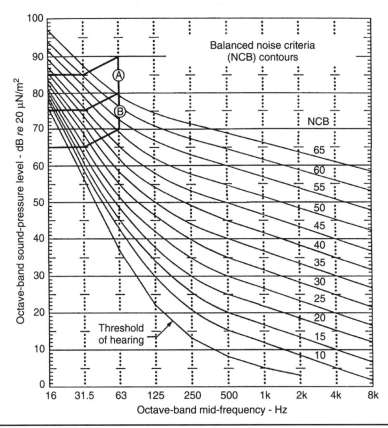

FIGURE 11-7 Standard NCB contour.

frequencies and because many noise sources are dominated by low-frequency content. An NCB-15 curve is very stringent and would be used in a critical application such as a recording studio. These curves are the balanced noise criteria (NCB) curves (Beranek, 1989), which, for most purposes, should be used instead of the older NC curves. The NCB curves are extended downward two octaves below NC to the octave band centered on 16 Hz. Another difference is the inclusion of the A and B areas. Sound-pressure levels of 70 or 80 dB magnitude below 63 Hz and which fall in the A and B areas might induce audible rattles or noticeable vibrations in lightweight partitions and ceiling structures. Other than these factors, use of the NCB and the NC curves is similar. Noise criteria curves can also be applied to HVAC design as discussed in Chap. 10.

During the noise survey, a sound-level meter is used to measure the sound-pressure level in each octave band. The noise is described by this series of sound-pressure-level measurements in octave bands centered at 16 Hz to 8 kHz. These sound-pressure levels are then plotted on the family of noise criteria curves, such as the black dots shown in Fig. 11-8. The noise level can then be described by the highest criteria curve touched by the dots. In this example, the sound-pressure level in the 1-kHz octave band touches the NCB-20 curve, and can thus be described in simplified terms as an NCB-20 noise level. In this way, different noise levels can readily be compared on a single-number basis.

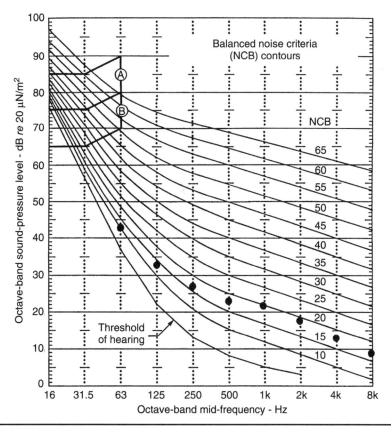

Figure 11-8 Use of NCB contours.

The use of NCB curves is described in *ANSI Standard 12.2 1995 (R1999), American National Standard Criteria for Evaluating Room Noise,* and in the *American Society of Heating, Refrigerating and Air-Conditioning Engineers (ASHRAE) Handbook.*

Relating Noise Measurements to Construction

The noise criteria contours can be used when designing the partitions of rooms, and in particular determining what transmission loss is required. For example, Fig. 11-9(A) shows an example of a plot of noise levels taken on site. If the design criterion is the NCB-20 contour, the intervening area between the curves shows the transmission loss required of the room partitions across the frequency spectrum. To obtain the required TL curve, the difference between the curves can be plotted at each standard frequency. In this example, at 500 Hz, the difference is 44 dB, and at 1 kHz, the difference is 40 dB, as shown in Fig. 11-9(B). With this curve, different types of partitions can be examined to find the one that is most suitable for meeting the isolation requirement. Ideally, the TL curve of the selected partition should exceed the difference curve at all frequencies. It is still possible that occasional exterior noises at the site may exceed the designed TL, and those noises will be audible inside.

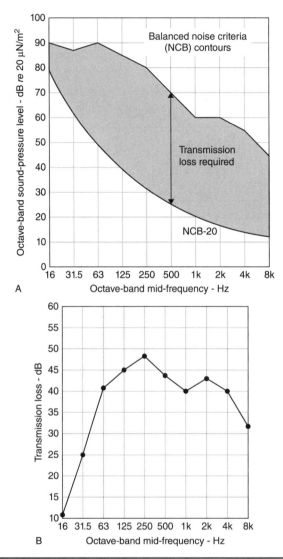

FIGURE 11-9 Use of site measurements and an NCB contour to determine required transmission loss. (A) The TL can be determined by plotting measured site noise levels (top curve) and the desired noise contour (NCB-20, the bottom curve). (B) The difference between the curves shows the required transmission loss.

The highest environmental noise levels are usually produced by trains, airplanes, and vehicular traffic. To prevent such noises from interfering with studio activities, partitions must be designed to reduce such noises to a tolerable level. While vehicular traffic noise may be fairly continuous, train and airplane noise may be occasional. Figure 11-10 shows the spectrum of jet noise as it passes over a studio. The lower curve is the NCB-15 contour, which is either the measured background noise of the studio or the studio noise goal, depending on the stage of the operation at which this study is made. The difference between the two curves, in dB, is the attenuation that the studio walls must provide.

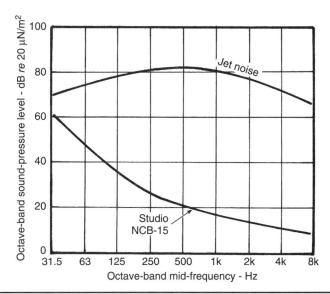

FIGURE 11-10   A comparison of measured airplane noise versus a design contour of NCB-15.

Figure 11-11 plots the difference between the jet noise curve and the studio noise curve of Fig. 11-10. This is the attenuation that the studio walls must deliver to bring the outside jet noise down to the desired background noise inside the studio.

What kind of construction is required to give the attenuation of Fig. 11-11? This is the subject of Chaps. 7, 8, and 9. The light, broken line in Fig. 11-11 is the Sound Transmission Class STC-60 contour. Walls rated as STC-60 (Chap. 7) are the more effective type. That class of walls offers approximately 60 dB of attenuation at 500 Hz. Without belaboring the subject, this is the general method of solving the internal background noise problems of studios and other sound-sensitive spaces. When designing specific structures, attention must be paid to factors such as building codes, materials, costs, and so forth.

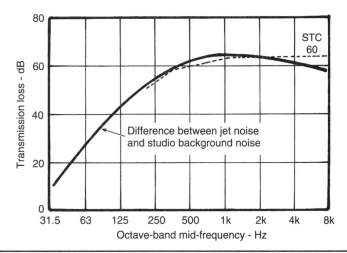

FIGURE 11-11   Difference between exterior and interior noise showing an STC-60 requirement.

Evaluation of Room Acoustics: Objective Methods

As with any endeavor, particularly ones as expensive as building acoustical spaces, there must be some way to evaluate the relative success or failure of the project. In other words, we must know if the room sounds good or bad. The inauguration of a new room usually generates as many opinions as there are listeners; it is important to have a systematic approach to evaluate and expertly summarize the acoustical performance of the room. As a baseline, many different objective measurements may be taken. For example, the room's pink-noise response, energy-time curve (ETC), and impulse response may be recorded at a variety of locations. These are valuable in characterizing a room. An example of an ETC plot is shown in Fig. 11-12. In this plot, the leftmost trace is the direct sound arriving at the microphone; the response to 250 msec shows early reflections; and the response after 250 msec shows reverberation.

The impulse response provides comprehensive information on a room's acoustical response. Moreover, after an impulse response has been recorded, it can be used to simulate the sound field of a room. An impulse response uses a single pulse to excite the acoustics of a room, and the response to the pulse is recorded. The impulse response is a unique signature of the room, and is different for each location where the pulse is sounded, and each location where the response is recorded; this is particularly true in large rooms. The pulse is a single loud sound such as gunshot. The frequency response of a short pulse encompasses all the audible frequencies. A plot of the impulse response shows the direct sound and the reflections from all the reflective surfaces in the room, as well as their amplitude and timing. When a room is acoustically altered, a new impulse response can be compared to the original response to evaluate the significance of the changes. In practice, when speech syllables or music notes are sounded in a room, they act as variations on a theoretical impulse as they excite the room's response. The room's impulse response can be convolved with an anechoic music recording to produce a simulation of how the music would sound when played in the room.

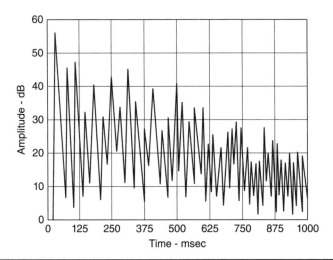

Figure 11-12   An example of an energy-time curve (ETC) measurement.

An impulse response measurement records sound pressure at the sensing microphone over time. From this measurement, several other metrics can be observed. Reverberation time (RT_{60}) measures the time it takes for reverberation to decay 60 dB. Early decay time (EDT_{10}) measures the time it takes for reverberation to decay 10 dB, and that time is multiplied by 6. Early-to-late energy ratios (EL_t) logarithmically compare early energy levels to late reverberant energy levels in the impulse response using a specified time as the cutoff between the two: $EL_t = 10\log$ (early energy)/(late energy). When the cutoff time t is set at 80 msec, the measure is termed C_{80} and is referred to as the clarity index or clearness index. Relative loudness (L) and relative strength (G) measure the sound level at a particular location compared to the sound level 10 meters from the sound source in an anechoic room. Support ratio (ST1) compares early sound reflected within a stage area to the direct sound; this allows musicians to properly hear each other and play in ensemble.

Evaluation of Room Acoustics: Subjective Methods

Subjective evaluation is also a very important component in a room's characterization. After all, ultimately it is the human impressions of an acoustical space that matter the most, not the numbers of a test instrument. Many methods have been devised to subjectively evaluate room acoustics; in some cases, these methods correlate objective measurements with subjective impressions. Various terms have been devised to describe acoustical performance. They can be categorized, for example, in three general groups: (1) Clarity (articulation, definition, intelligibility); (2) Ambience (spaciousness, reverberance, tonal quality); (3) Loudness. These criteria are interrelated and are mainly defined by the level, timing and direction of reflected sounds relative to the direct sound.

In many cases, the evaluation criteria are greatly influenced by events such as early reflections that occur soon after the arrival of the direct sound. In particular, much of a room's music acoustical quality is determined within 80 msec, and is reinforced or degraded by later events. As noted, a measure for music clarity using the 80-msec time is specified as C_{80}. For speech, the timing is often reduced to 50 msec; a measure for speech clarity is specified as C_{50}. Clarity can also be rated using a number of objective measurements such as the speech transmission index (STI), rapid speech transmission index (RASTI), articulation loss of consonants (%Alcons), and early-to-late reverberation ratio (ELR). STI measures the intelligibility of speech in the presence of reverberant energy and background noise; RASTI is a simplified version of STI; %Alcons measures the perception of spoken consonants; ELR compares the energy of early reflections to that of late reflections. The relationship between some of these speech measures is shown in Table 11-4.

Ambience is most closely associated with spaciousness and reverberance. Spaciousness can be quantified as the interaural cross-correlation coefficient (IACC), lateral

% Alcons	100	57.7	33.6	19.5	11.4	6.6	3.8	2.2	1.3	0
STI	0.1	0.2	0.3	0.4	0.5	0.6	0.7	0.8	0.9	1.0
C_{50} (dB)	−12	−9	−6	−3	0	3	6	9	12	15
Subjective Rating	Poor			Fair			Good			

TABLE 11-4   Approximate Relationships between Ratings of Speech Intelligibility

energy fraction (LEF), apparent source width (ASW), and listener envelopment (LEV). IACC measures the dissimilarity of signals at the listener's left and right ears; LEF is the ratio of lateral energy to omnidirectional energy over the initial 80-msec period; ASW measures the stereo effect and perceived size of the source such as from early lateral reflections; LEV measures the perception of the surround sound in a room such as from late lateral reflections.

Reverberance is often measured by reverberation time (RT_{60}), early decay time (EDT), and early-to-late reverberation ratio (ELR). RT_{60} measures the full extent of reverberation over time; EDT measures liveliness of short-term reverberation; as noted, ELR compares the energy of early reflections to that of late reflections. In many ways, reverberance is contrary to clarity; in a good room design, the two must be balanced. This explains why a room that has good speech acoustics is probably poor for most music, and vice versa. Some room designs attempt to reconcile clarity and reverberance with a relatively quickly decaying early reverberation followed by a more slowly decaying late reverberation; this allows clarity, but also provides a sense of spaciousness.

Loudness is certainly one aspect of a room's acoustical performance. But, it is relatively fixed depending on the size of the room, and loudness of the sound source. For example, when the source is an orchestra and the hall is the size of a concert hall, the loudness of the orchestra in the hall is not a design variable. In a speech auditorium, other factors such as intelligibility clearly take precedence over loudness; if the level of natural speech is too low, then amplification is needed.

Leo Beranek notably designed numerous concert halls and devised ways to evaluate their acoustics. He consolidated many years of experience in acoustical evaluation into seven criteria: reverberance, loudness, spaciousness, clarity, intimacy, warmth, and hearing on stage (Beranek, 1992). The evaluation of intimacy (or presence) is based on the time difference between the direct sound at the listening position and the first reflected sound. Beranek refers to this as the initial time-delay gap (ITDG) and estimates that the gap should be 20 msec or less. This value can be achieved, for example, in a traditional shoebox-shaped hall, or by placing niches with enclosing walls, balconies or other sound reflectors near the main listening areas. The early reflections with a short initial time-delay gap also give an impression of spaciousness.

For accuracy, the terms used in a subjective evaluation must be carefully defined; if different listeners have different understandings of what the subjective evaluation terms mean, then their evaluations are necessarily inconsistent and the results flawed. Moreover, to benefit from the subjective impressions of the room acoustics, the terms must be correlated to the acoustical properties of the room that help determine them. For example, intimacy is a perceived closeness of the orchestra; it is usually achieved when the initial-time-delay gap is less than 20 msec.

Warmth is a feeling of a dark and rich sound as opposed to a too-bright sound; it is achieved when the reverberation time of bass frequencies is longer than the reverberation time of mid and treble frequencies. Warmth is a good example of an evaluation term that can be determined subjectively, and also measured analytically. In particular, the bass ratio is the sum of the reverberation times at 125 and 250 Hz divided by the sum of the reverberation times at 500 and 1000 Hz. A bass ratio greater than 1.0 would indicate a subjectively warm concert hall. Figure 11-13 is a plot of reverberation time versus frequency; it shows an increase (1.5×) in reverberation time at frequencies below 500 Hz. The bass ratio quantifies warmth. Similarly, the treble ratio is the sum of

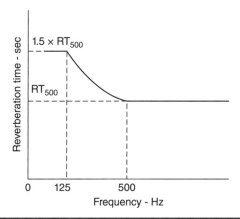

Figure 11-13   The bass ratio quantifies warmth in a room; an increase in reverberation time at low frequencies indicates warmth.

the reverberation times at 2 and 4 kHz divided by the sum of the reverberation times at 500 and 1000 Hz. A treble ratio greater than 1.0 would indicate a subjectively brilliant concert hall.

To organize the listener's subjective impressions, an evaluation scoring form may be used; examples of two forms are shown in Figs. 11-14 and 11-15. The evaluation is completed as music (or other type of presentation) is performed in the room. Because room acoustics can vary widely in different locations in a room, the listener uses a separate form for every seat used for evaluation. The better the room, the less the variation in acoustics throughout the room. The listener marks specific aspects of the acoustics such as warm bass, loudness, and diffusion. The listener may also record an overall impression, for example, using A, B, C, D, and F grades. Evaluations should be undertaken when the rooms are occupied with people and being used as intended. For example, a concert hall's acoustics will usually differ dramatically depending on whether the hall is empty or full.

Table 11-5 summarizes many of the acoustical qualities that are analyzed when evaluating the acoustical performance of a room. The table also lists the definitions of the qualities, architectural features that are thought to influence them, and objective measures. This table uses concert halls as its basis, but the evaluation terms can be applied to many other kinds of acoustical spaces.

Articulation Index and the Spoken Word

The Articulation Index (AI) can be used to gauge the speech intelligibility in an auditorium, place of worship, or other room where speech or dialogue is prominent. AI is subjectively evaluated. A series of words and sentences are spoken, and a group of trained listeners write down what they hear. The index nonlinearly relates to the percentage of correctly heard responses, and it differs between words and sentences. An AI of more than 0.7 is considered to be very good (90 correct word responses out of 100). In a hall, some seating areas may achieve a very good AI, while some seating areas may

Room Acoustics Evaluation Form

Hall: _____ Position: _____

	1	2	3	4	5	6	7	O
Clarity	Not clear enough	2	3	4	5	6	Extremely clear	Cannot tell
Intimacy	Not intimate enough	2	3	4	5	6	Extremely intimate	Cannot tell
Envelopment	Not surrounding enough	2	3	4	5	6	Extremely surrounding	Cannot tell
Balance	Too much bass	2	3	Well balanced	5	6	Too much treble	Cannot tell
Reverberance	Not reverberant enough	2	3	4	5	6	Too reverberant	Cannot tell
Loudness	Not loud enough	2	3	4	5	6	Too loud	Cannot tell
Overall impression	Very bad	2	3	4	5	6	Very good	Cannot tell
Background noise	Not audible	2	3	4	5	6	Too loud	Cannot tell
Echoes	None detected	2	3	4	5	6	Clearly heard	Cannot tell

Comments:

FIGURE 11-14   An example of a form used to subjectively evaluate room acoustics. (*from Cervone, 1990*)

not; it is the responsibility of the acoustician to provide uniformly high AI throughout the hall. In some buildings such as offices, a low AI is desired so that nearby speech is unintelligible and thus less distracting and more private. As an alternative to AI, the Rapid Speech Transmission Index (RASTI), an objective test method, can be used to measure speech intelligibility.

Places of worship can present unique acoustical challenges. A religious service may feature spoken word, chant, individual singing, congregational singing, musical performance, pipe organ, or combinations of any of these. Because of the diversity of the

Room Acoustics Evaluation Form

Venue: _____ Date: _____

Seating capacity: _____ Volume (ft^3): _____

Seat number: _____ Seat location: _____

Type of music performance: _____

_____

Overall impression: _____

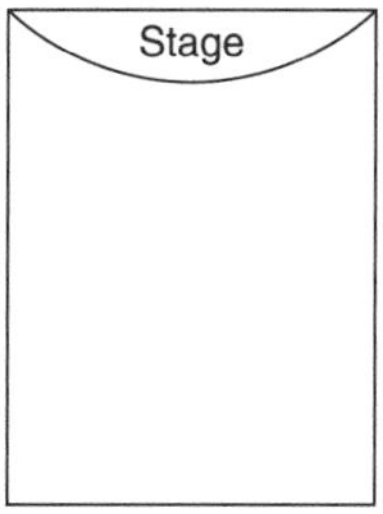

Stage

Clear sound |_____|_____|_____|_____| Blurred sound

(varies from clear or distinct to blurred or muddy)

Live reverberance |_____|_____|_____|_____| Dead reverberance

[liveness or persistence of mid-frequency sound]

Warm bass |_____|_____|_____|_____| Cold bass

(relative liveness of bass or longer duration of reverberance of bass compared to mid- and treble frequencies)

Intimate sound |_____|_____|_____|_____| Remote sound

(auditory impression of apparent closeness of orchestra)

Satisfactory loudness |_____|_____|_____|_____| Unsatisfactory loudness

[indicate early or direct sound (D) and reverberant sound (R) on scale]

Good diffusion |_____|_____|_____|_____| Poor diffusion
(expansive sound)   (constricted sound)
(envelopment of sound which surrounds listener from many directions)

Good balance |_____|_____|_____|_____| Poor balance

(between soloist or chorus and orchestra, among sections of orchestra)

Satisfactory |_____|_____|_____|_____| Unsatisfactory
background noise   background noise
(very quiet)   (very noisy)

(from HVAC, or intruding noise inside or out)

Echoes |____| No |____| Yes Direction: _____

(long delayed reflections that are clearly heard)

Note: Use separate form for each seat where acoustics are evaluated.

Figure 11-15 An example of a form used to subjectively evaluate room acoustics. (*from Egan, 1972*)

Acoustical Quality	Definition	Architectural Feature	Acoustical Measure
Envelopment and source width	Early reflections arriving at the listener from the side in the first 80 msec after direct sound	Narrow rooms (70–80 ft) or multiple tiers of narrow balconies or niches	Lateral energy fraction (LEF)
Clarity	Reflections arriving soon after direct sound	Reflecting ceiling, canopy or panels	Clarity index (C_{80}) and early-to-late energy ratio (EL_t)
Reverberance	Persistence of sound with smooth decay	Large room volume, reflecting materials, reverberation chamber	Reverberation time (RT_{60})
Loudness	Reflections arriving soon after direct sound	Room size, reflections from ceiling and walls, proximity to source, sight lines	Relative loudness (L) or relative strength (G)
Intimacy	First reflections arriving soon after direct sound	Reflections with short path lengths	Initial time-delay gap (ITD)
Warmth	Persistence of low frequencies and low-frequency reverberation	Heavy and massive building materials	Bass ratio
Brilliance	Persistence of high frequencies and high-frequency reverberation	Reflecting surfaces	Treble ratio
Spaciousness	Dense late sound energy (after 80 msec) arriving from sides	Large room volume, sound diffusing materials	Interaural cross correlation (IACC)
Localization	Strong direct sound relative to reflected sound	Clear sound and sight lines, proximity to source	Early loudness level
Ensemble	Sound reflections that allow musicians to hear each other across a stage	Overhead and side wall reflecting surfaces at performance area	Support ratio (ST1)

(*from Siebein and Gold, 1997*)

TABLE 11-5 Qualities Used to Evaluate the Acoustical Performance of Rooms

presentation, tuning the absorption and reflection in the room can be difficult. For example, if the reverberation time is too long, spoken word may be unintelligible. If the reverberation time is too short, music may sound unnatural. It is important to supply adequate reverberation in the congregation seating area; if that area is acoustically dry, it may give individuals the feeling of singing or speaking alone, rather than as part of the shared acoustical space. Room details, such as a canopy over a pulpit that reflects and reinforces sound to the seating area, can solve acoustical problems. In some architectural designs, variable acoustics are used to optimize acoustical performance for different applications.

In many cases, because an acoustical balance is difficult to supply naturally, electronic means are used. For example, digital reverberation could be used to supplement the sound of a pipe organ in an acoustically dry space, or a sound-reinforcement system could be used to improve intelligibility in a reverberant space. Very generally, a smaller room volume-to-seat ratio is better for spoken word, and a longer ratio is better for music performance.

Traditional church plans, using natural acoustics, often use narrow floor plans and high ceilings. This plan provides a feeling of intimacy and open space, because of short initial time-delay gaps from the side walls, and reverberation from the high ceiling. Churches with a wide floor plan, in which the congregation is seated around a central pulpit or stage, may have longer time-delay gaps, which may decrease the sense of intimacy; sound reinforcement may be needed.

Room Reflections and Psychoacoustics

The importance of reflected sound in concert halls has been known for many years. In particular, the contribution of lateral reflections to the enjoyment of the music is recognized to the extent that designers of concert halls must take special care that there be an adequate supply of such reflections (Beranek, 1962). There has been considerable research on the perception of reflected sound. In a typical experimental setup, one loudspeaker, carrying the direct sound, is pointed at the listener. To one side, another loudspeaker emits the simulated delayed reflection. Many observations with simulated reflections of various amplitudes and time delay provide the basic data from which conclusions can be drawn.

Figure 11-16 shows the effect of lateral reflections on speech signals. Below the threshold curve A, reflections are inaudible. Above curve C, reflections have deteriorated to undesirable, discrete echoes. The space between curves A and C is a useful area that defines how lateral reflections can affect the sound in practical listening situations.

The data of curve A was obtained by Olive and Toole (Olive and Toole, 1989). Using speech as the signal, the listener heard the direct sound with a simulated reflection of variable amplitude and delay superimposed. The listener was instructed to describe exactly what was heard. The listener hears no trace of a reflected sound until its amplitude and delay reached curve A. This curve is called the absolute threshold of reflection audibility. In the shaded area below curve A, no reflections are audible. Curve C defines the conditions under which the reflections are perceived as discrete echoes; this curve is a composite from data of other researchers (Lockner and Burger, 1958; Meyer, and Schodder 1952).

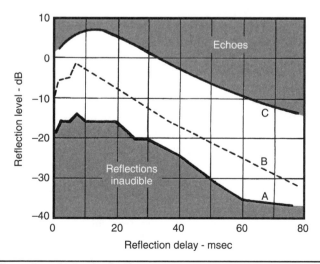

Figure 11-16 Perception of lateral reflections.

When the amplitude of reflections was about 10 dB above the threshold curve A, listeners noted that something different was happening to the direct signal. In addition to spaciousness, shifting and spreading of the front auditory image were noted, and this contour has been designated as curve B. Spaciousness was sensed both below and above curve B.

It is remarkable that even though these tests were run in an anechoic room, the room was given a sense of spaciousness, of being much larger and of a different character. The listeners were not aware of the reflection as a separate event; the sound of the front (direct) loudspeaker was given this spacious character by the presence of the reflected component. This benefit underlies the importance of lateral reflections in hall design.

Spaciousness is a characteristic highly prized in concert halls, which are designed to supply the lateral reflections necessary to create the impression of spaciousness. Now that such reflections have been well defined (as in Fig. 11-16), the possibility arises of applying this knowledge to a listening room. By carefully adjusting the amplitude and delay of early reflections, a spacious sense can be given to the loudspeaker sound, the room can be made to seem larger, and the stereo image can be optimized in regard to spreading of the image. Listening room design is considered in subsequent chapters.

Room Modeling

Computer software is often used to design architectural spaces, and predict and simulate their acoustical performance. These programs allow the acoustician to "build" the spaces in software and then simulate the sound field that will exist in the completed physical structure. Based on the analysis of the model, the acoustical design can be altered to provide optimal acoustics. In some cases, beyond numerical prediction, the modeled sound field can be aurally simulated and evaluated through speakers or headphones. The virtual impulse response can be convolved with anechoically recorded music to simulate how music or speech will sound in the finished room.

In computer-based models, spaces are built in three dimensions; in some cases, the acoustical analysis program can use designs generated by architectural CAD programs. Virtual sound sources and microphones are placed and the impulse response and other standard testing can be performed. Ray tracing and imaging may be employed. In ray tracing, sound rays from a source strike the reflective surfaces in a room and their arrival times and amplitudes at a location are recorded. The reflective surfaces are modeled with the appropriate absorption coefficients, and diffusion is accounted for. With the imaging method, the program identifies the reflective surfaces and creates ray paths to the source.

Some acousticians prefer to construct an actual physical model of an interior space. The model is built to scale (for example, 1:96, 1:48, or 1:10 for smaller spaces). Small sound sources and microphones are used to study acoustical performance; the frequency of the test signals must be scaled up to correspond to the model. For example, in a 1:10 model, a frequency of 5,000 Hz is used to predict response in the full-size room at 500 Hz. The model is built using materials that possess the correct absorptive and reflective characteristics corresponding to the full-size room; for example, in a 1:10 model, absorption of a modeling material at 5,000 Hz must be equal to the absorption of the building material at 500 Hz. A spark may be used as an impulsive sound source. At high modeling frequencies, air absorption must be accounted for.

CHAPTER 12

Recording Studio for Pop Music

A studio designed for recording popular music ranks among the most ambitious of architectural and acoustical projects. Large professional studios can cost well over a million dollars before equipment costs are considered. Smaller studios cost less, but are still relatively expensive structures. To facilitate the work done there, pop studios usually contain several very specific acoustical features such as a drum and vocal booths. A studio such as this can be used to record a wide variety of music such as pop, rock, hip-hop, rap, R&B, country, soul, and jazz.

Any recording studio, large or small, is critical in terms of acoustical requirements. This is the place where music performances are recorded for preservation. Along with the music, anomalies and defects in the studio's acoustics are also unfortunately captured and replayed countless times. Thus the acoustical treatment of a studio must be carefully designed and constructed so that the acoustics are pleasing and complementary to the music, and also relatively neutral so the acoustics do not intrude on the music.

The reverberation time in a pop studio should be relatively short. For example, a broadband reverberation time of less than 0.4 sec is typical. The short reverberation time yields relatively dead tracks particularly when close micing is used. It also provides better isolation between simultaneously recorded tracks. This makes it easy to add digital reverberation and other effects to individual tracks during the mixdown. Proper absorption is thus a key element of the studio design. Too little absorption makes the studio too live. Too much absorption is problematic because musicians need some reverberation for comfortable performance.

The high sound-pressure levels typical in a pop studio (and its control room) must be controlled. High levels in the studio that are audible in the control room can adversely affect decisions made during tracking; clearly, ideally, only the control room monitors should be heard. Conversely, loud monitor levels in the control room could be picked up by microphones in the studio. Thus the wall between the studio and control room, as well as walls between different recording spaces in the studio, must provide good isolation. Although not as critical, some isolation is needed between the main studio space and other rooms such as drum and vocal booths.

In addition, precautions must be taken to avoid disturbing neighbors outside the studio complex. Containing the high-level sounds of pop music is a very real technical challenge. If the sound levels in the studio or control room reach 120 dB (near the threshold of pain), a 50-dB wall is needed to reduce music levels to a tolerable 70 dB. For this reason, providing isolation around the studio is a high priority in the proposed studio design presented here.

Design Criteria

For this studio design example, we shall assume that the space to be used for the studio is on the ground floor of an existing concrete-structure building. It is a corner space in this example; the studio has two exterior walls and two interior walls, and there are neighbors on the same floor and on the floor above. The isolation criteria of the floor and walls are carefully considered; to help maintain isolation between rooms, a sound lock is included. In addition, treatment is designed to provide a suitable reverberation time.

The control room is acknowledged only by placement of an observation window in the south wall. Control room design is a separate project and will be carried no further in this chapter. Control rooms are considered in Chap. 15. To maintain low background noise levels and to minimize passage of sound between rooms, the HVAC design is very important; HVAC is discussed in Chap. 10.

Floor Plan

The general floor plan is shown in Fig. 12-1. The facility includes a drum booth, a vocal isolation booth, an equipment storage room, a baffle storage area, and a sound lock that opens into the control room as well as the exterior. A control room window is included in the south wall. Approximately half of the studio floor area is carpeted, with wood parquet on the other half. Different floor coverings help provide different acoustical characteristics and thus more variability when recording. The volume of the studio is about 15,220 ft^3.

Wall Sections

Figure 12-2 locates sections D-D, E-E, F-F, and G-G (sections A-A, B-B, and C-C are reserved for drum-booth walls, which are discussed later). Sections D-D and G-G are exterior walls; E-E and F-F are interior walls. The inside walls for the drum booth, vocal booth, and equipment room are not indicated pictorially but comprise simple single layers of 5/8-in gypsum board on each side of metal channels. Using Fig. 12-2 as a guide, each of the wall sections can be examined individually.

Section D-D

As noted, isolation is a critical element in this studio design. Thus a floating floor is incorporated in the design. Sections D-D and E-E are both shown in Fig. 12-3(A). The floor (5) of 4-in concrete, common to both, is floated on compressed glass fiber or neoprene pads. The concrete is not poured until the 3/4-in plywood form is in place and covered with a plastic membrane to prevent concrete from seeping through cracks and

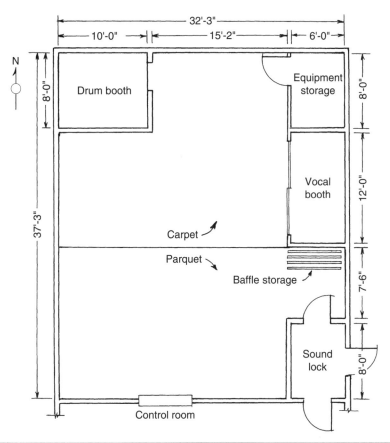

Figure 12-1 Plan view of a pop music recording studio.

forming a solid bridge to the supporting structure. Glass-fiber perimeter boards separating the floating floor from the walls are also to be shielded by plastic membrane. The floating concrete floor must be completely isolated from the structure to be effective. The pads must be selected and sufficient in number to bear the weight of the concrete floor itself, the walls, and the ceiling, and be deflected about 15%, so that the full resiliency of the pads can be realized.

For optimal performance, the floor's mass and static deflection of the pads must be carefully selected; in particular, the floor's resonant frequencies must be calculated. As a lower cost alternative, a wood floating floor may be used. Floating floors are discussed further in Chap. 8.

All the interior walls of the recording studio are supported on the floating floor. The upper ceiling (1), hung from the structural building on resilient isolation hangers (6), is made of two 5/8-in gypsum-board layers separated by a septum of dense vinyl. This is used to increase the acoustical isolation to occupants on the second floor. The floating floor, the four walls, and this acoustical ceiling constitute a space within a space for attenuation of the music within.

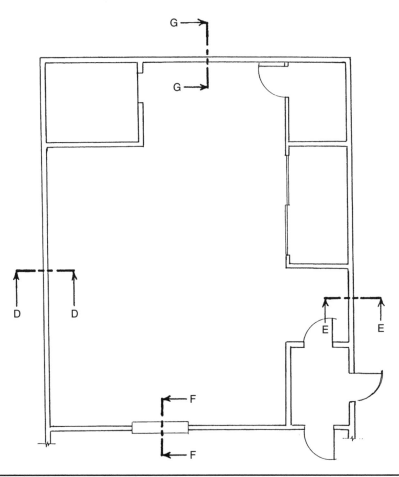

Figure 12-2 Locations of four sectional views of a pop music studio.

The lower acoustical ceiling (2) uses a standard suspended T-frame ceiling with a 16-in airspace to conform to the standard C-400 (400-mm) mounting. The ceiling uses 1 × 24 × 24-in Tectum ceiling tiles laid in the T-frame with 6-in of glass-fiber backing resting on the top of the tiles. The section also shows wall stabilizers (9) designed to isolate the walls from the structure.

A spacing wall (3) is built alongside the concrete west wall of the building with a 2-in spacing. This spacing is loosely filled with 1-1/2-in glass fiber. The layer of the new wall is 5/8-in gypsum board attached to steel studs and channels. The inner space of the wall is filled with 4 in of low-density building insulation and the inner layer is a double layer of 5/8-in gypsum board. As with any acoustically sensitive construction, all cracks and joints should be filled with acoustical sealant, the goal being to make the inner space hermetically independent. Cracks that allow air passage will also allow sound to pass through the barrier, defeating the purpose of the carefully designed walls, floor, and ceiling. All metal wall runners should be set in acoustical sealant.

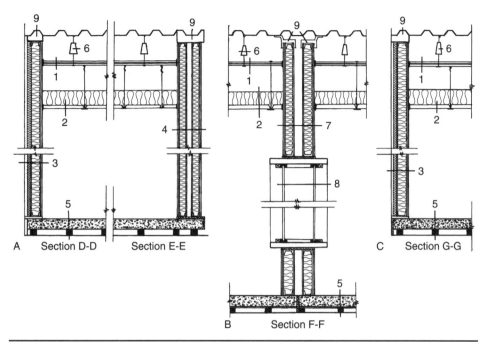

Figure 12-3 Wall sections of a pop music studio. (A) Sections D-D and E-E, exterior and interior walls. (B) Section F-F, interior wall between studio and control room. (C) Section G-G, exterior wall.

For common sense to be kept abreast of theory, it should be emphasized that resilient hangers (6), floating floor (5), and spacing wall (3) from the wall of the building are all employed to isolate the studio space from the structure. The reason is that the concrete structure is an excellent transmitter of noises both into and out of the studio. The primary justification for this expense is to isolate the neighbors from the studio.

Section E-E

The only feature of section E-E [see Fig. 12-3(A)] not discussed in terms of section D-D is double wall (4); it is an inner wall that runs the length of the studio. It is a double wall spaced 2 in. Each leaf of the double wall is composed of standard metal studs and channels with a single 5/8-in gypsum-board layer on the inside and a double layer of the same on the outside. Loosely stuffed 4-in glass fiber is placed inside the walls. Although not specified, it would be advantageous to increase the STC isolation by a few decibels by mounting the outer layer of 5/8-in gypsum board facing the other occupants on resilient channels.

Sections F-F and G-G

Section F-F is shown in Fig. 12-3(B). This shows the double wall (7) that separates the studio from the control room. It must provide high transmission loss so that sound levels from the studio do not affect monitoring in the control room, and conversely so that monitoring levels do not leak into the open microphones in the studio. The double wall is the same as wall (4) shown in section E-E, but with 6-in spacing.

The wall incorporates a double-pane glass observation window (8). Because most windows usually have lower transmission loss than the surrounding walls, care must be taken in the window design so that it does not compromise the isolation integrity of the wall. The window uses two heavy (or laminated) panes of different thicknesses, each isolated from the frame by rubber edging. The frame, in turn, is isolated from the wall structure by pads of compressed glass fiber. Window design is discussed further in Chap. 9. Each pane is mounted in a separate wall built on separated slabs. In some designs, a third interior load-bearing wall is used; alternatively pipe columns can be used for support. If a third glass pane is added, this can mildly improve high-frequency insulation; however, care must be taken to ensure that low-frequency insulation is not diminished.

The surfaces inside the two window panes are lined with open-cell foam or glass-fiber boards. In some designs, this cavity doubles as a Helmholtz resonator. If desired, the window panes can be angled downward; this may usefully deflect sound downward and prevent an unwanted reflection in the room. Moreover, the angle can reduce light reflections. In addition, the angle moves the upper wall sections farther apart; among other thing, this can provide room for soffit-mounted monitor loudspeakers. When designing soffit mounts for loudspeakers, care must be taken to ensure that the monitors are properly insulated, and sound will not enter the studio.

Section G-G is shown in Fig. 12-3(C). All features of this section have been previously discussed.

Drum Booth

The purpose of a drum booth is to reduce the high sound levels from the drum kit at the positions of the other instruments (and microphones) in the studio, and to provide a relatively dead local environment for recording the drums. An isometric sketch of the proposed drum booth is shown in Fig. 12-4 and details of its construction are given in Fig. 12-5. The drum booth measures $8 \times 10 \times 11$ ft with openings on the south and east sides to give a clear view of the studio. The booth features a wood floor of 1-1/2-in tongue and groove decking on top of a 2×8-in structure deadened by loading with sand. The entire floor is isolated from the building structure by compressed glass fiber or neoprene pads such as used to float the floor. This gives the drummer the required solid floor which is essentially nonresonant. If desired, for improved insulation, double-pane glass windows can be placed in some portions of the openings. The entryway to the booth can remain open; alternatively, for better isolation, a solid door or sliding glass door could be installed.

To deaden the ambient sound field near the drums, the ceiling of the booth is made highly absorbent by packing the 2×12-in frame with glass-fiber building insulation. The north and west walls of the booth are reflecting surfaces of 3/4-in plywood, but made into a tuned absorptive structure with 1/2-in holes spaced 7 in. These two walls have a tuned absorption peaking in the region of 80 Hz, but little absorption above 150 Hz.

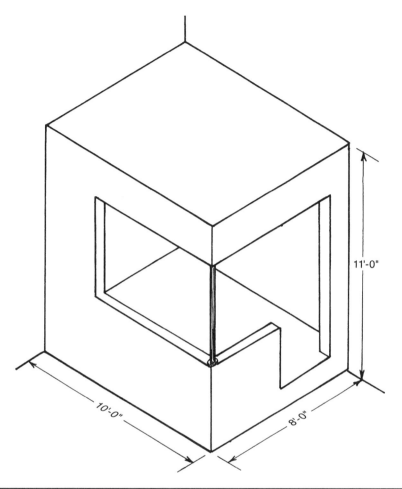

Figure 12-4   An isometric sketch of a drum booth.

This absorption helps deaden the kick drum, yet provide the drummer with adequate personal return.

Vocal Booth

The vocal booth is a 6 × 12-ft space with a sliding glass door opening to the studio, which provides entry as well as an excellent view both ways. This sliding glass door, when closed, will provide about 20 to 25 dB of insulation in addition to the inverse-square (or "6 dB/distance-double") law. The interior of the vocal booth should be treated with absorptive material similar to that used to treat the walls of the announce booth described in Chap. 16. In addition, treatment should be added

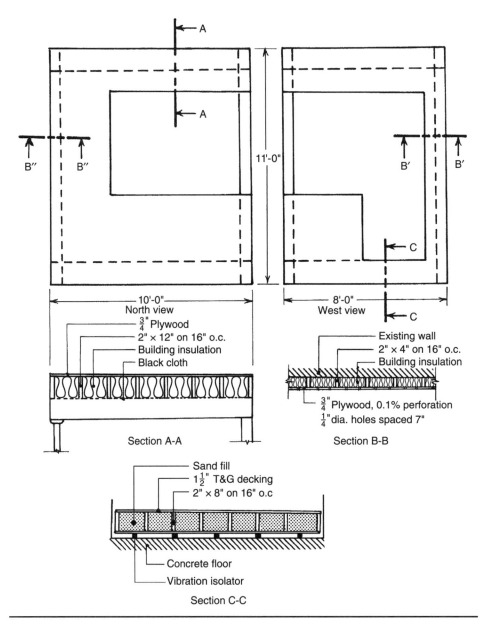

A

A

11'-0"

B″ | B″

B′ | B′

C

C

10'-0"
North view

8'-0"
West view

$\frac{3}{4}$" Plywood
2" × 12" on 16" o.c.
Building insulation
Black cloth

Section A-A

Existing wall
2" × 4" on 16" o.c.
Building insulation

$\frac{3}{4}$" Plywood, 0.1% perforation
$\frac{1}{4}$" dia. holes spaced 7"

Section B-B

Sand fill
$1\frac{1}{2}$" T&G decking
2" × 8" on 16" o.c

Concrete floor

Vibration isolator

Section C-C

FIGURE 12-5 Drum booth design details.

to control axial modes. Absorption such as from bass traps in the corners would be best for this.

Studio Treatment: North Wall

The north wall of the studio is covered with 2-in Tectum panel with C-40 mounting, as indicated in Fig. 12-6. With this mounting system, the panels are furred out from the wall on 2 × 2-in strips (net about 1-1/2 in) on 24-in centers. This modest spacing from the wall improves the absorption of the panel so that it is highly absorptive above 250 Hz.

Studio Treatment: South Wall

A recording studio can be excellent acoustically, and very poor aesthetically. Some might associate drabness with poor sound quality. Musicians are aesthetes, and it is a fair guess that they, their guests, and clients would appreciate one wall of the studio devoted to an artistic piece. For this reason, the south wall of the studio (see Fig. 12-6) would be an

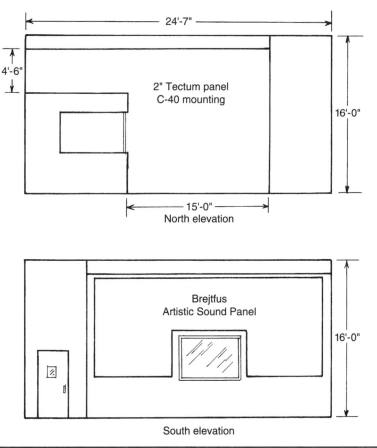

FIGURE 12-6 North and south interior walls of a pop music studio.

ideal place for a decorative absorptive panel such as the Brejtfus Artistic Sound Panel. Whatever the artistic design, companies such as Brejtfus not only can make it, they can make it absorptive. Their Artistic Sound Panel comprises a glass-fiber base on which a polyfoam layer bears the design. Different thicknesses and sizes are available; NRC is greater than 0.9. Clearly, if budget or taste does not allow an artistic treatment, low-cost absorption can be provided with glass-fiber panels such as described in Chap. 3.

Studio Treatment: East Wall

Broken up by many doors and other openings, the east wall can only accommodate an absorptive panel measuring 4 ft to 6 ft in width running almost the length of the studio. This is shown in Fig. 12-7.

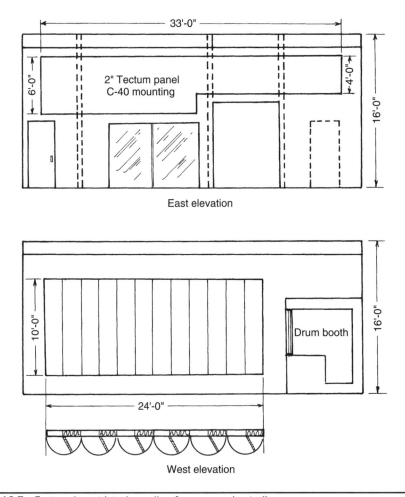

FIGURE 12-7   East and west interior walls of a pop music studio.

Studio Treatment: West Wall

An array of six doors, 10-ft high and 24-in wide, covers 24 ft of the west wall (see Fig. 12-7). Each door is hinged, and is absorbent on its inner side and reflective on the outer side. When each door is open, it not only exposes its own absorbent backside, but also reveals another absorbent area the size of the door. When all doors are swung open, the entire 10×24-ft area is absorbent.

The ultimate purpose of the doors is not to alter the reverberation time, because, as we will see, this 240-ft^2 area has only a minor effect on the overall reverberation time. Rather, the doors make available reflective and absorbent areas so that musicians can set up near one or the other to achieve the acoustical ambience they desire. Note that the Tectum panel wall on the north end of the studio is absorbent, but this area of swinging panels is the only other absorbent area close to musicians.

Studio Treatment: Floor and Ceiling

The floor is about half carpet and half parquet wood (see Fig. 12-1). A curved line of demarcation between the two areas would look better, but the style should be drawn by the occupants-to-be. The suspended ceiling is a very absorbent surface, even if the space behind is shared with air-conditioning ducts and electrical equipment.

Sound-Lock Corridor

The studio's floor plan (see Fig. 12-1) shows a sound-lock corridor. The inner door leads to the studio area, a second door leads to the control room, and a third door leads to the exterior. The purpose of a sound lock is clear; some insulation is provided even when one door is opened. When both doors are closed, the sound lock provides insulation well above that of a single door. To provide this additional insulation, the interior of the sound lock should be absorptive. Although studio wiring is outside the scope of this book, it may be advisable to place a microphone/headphone panel in a wall of the sound lock; this corridor can provide a uniquely dead recording space. Sound locks are discussed further in Chap. 9.

Reverberation Time

Multichannel recording demands close microphones or direct feeds to achieve sufficient channel separation. To assist close micing, a pop studio should present a short reverberation time. Moreover, reverberation-time calculations are of interest to balance the absorption of different areas and materials. In this pop recording studio design, almost every available ceiling and wall area has been covered with absorbent material. In other words, it has been made quite dead for a room of 15,220-ft^3 volume. Reverberation-time calculations also allow us to inspect the distribution of absorption about the room. Table 12-1 shows all the reverberation-time calculations for each absorbent unit contributing to the overall acoustics of the space.

For each of the six standard frequencies the absorption coefficient α for a given material is multiplied by the area S of that material to obtain the $S\alpha$ product in sabins. Summing all these unit absorbances in sabins for a given frequency gives the total

Description	Area, ft²	125 Hz		250 Hz		500 Hz		1 kHz		2 kHz		4 kHz	
		α	Sα	α	Sα	α	Sα	α	Sα	α	Sα	α	Sα
Suspended ceiling C-400 (16 in) Tectum ceiling tile 1 × 24 × 24-in lay-in 6-in glass-fiber backing	956	1.01	965	0.89	850	1.06	1013	0.97	927	0.93	897	1.13	1680
East-wall panel: Tectum wall panel C-40 mounting, 1-1/2 in	164	0.42	69	0.89	146	1.19	195	0.85	139	1.08	177	0.94	154
West-wall adjustable panels:													
Fully opened	240	1.0	240	1.0	240	1.0	240	1.0	240	1.0	240	1.0	240
Fully closed	0		0		0		0		0		0		0
South-wall panel: Brejtfus Artistic Sound Panel	210	0.16	34	0.47	99	1.10	231	1.14	239	1.05	221	1.04	218
North wall fully covered with Tectum wall panel C-40 mounting, 1-1/2 in	252	0.42	106	0.89	224	1.19	300	0.85	357	1.08	272	0.94	237
Drum booth ceiling	80	1.0	80	1.0	80	1.0	80	1.0	80	1.0	80	1.0	80
North and West walls 0.1% perforated 4-in deep	128	0.8	102	0.3	38	0.2	26	0.15	19	0.15	19	0.1	13
Floor, heavy carpet	448	0.08	36	0.24	108	0.57	255	0.69	309	0.71	318	0.73	327
Floor, parquet	465	0.02	9	0.03	14	0.03	14	0.03	14	0.03	14	0.02	14
Total absorption, sabins			1641		1799		2354		2324		2238		2365
Reverberation time with West doors open, sec			0.45		0.41		0.32		0.32		0.33		0.32
Reverberation time with West doors closed, sec			0.53		0.48		0.35		0.36		0.37		0.35

TABLE 12-1   Reverberation-Time Calculations for a Pop Recording Studio

absorption A for that frequency from which the reverberation time may be estimated from the equation:

$$RT_{60} = 0.049V/A \qquad (12\text{-}1)$$

where RT_{60} = reverberation time, sec
V = volume of room, ft^3
$A = S\alpha$ = total absorption of room, sabins

Given that $V = 15,220$ ft^3, and using the surface areas S and absorption coefficients α of the surface treatments, we can use the equation to calculate RT_{60} at different frequencies. The calculated reverberation times have been plotted in Fig. 12-8 for the swinging panels both open and closed. These panels will probably never be moved for the purpose of adjusting reverberation time. It is not possible to sense the difference between the 500-Hz reverberation times of 0.32 and 0.35 sec brought about by the swinging panels. Rather, the adjustment of these panels is primarily to adjust local acoustics to suit individual musicians. A musician can easily sense playing alongside a hard versus a soft wall.

Baffles

The traditional baffle (sometimes called a "gobo") is used to improve the separation between instruments. It typically has one reflective side and one absorptive side. The baffle suggested in Fig. 12-9 introduces diffusion into baffle technology. Sound returned

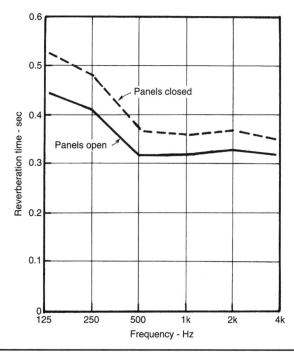

FIGURE 12-8 A reverberation time plot of a pop music studio.

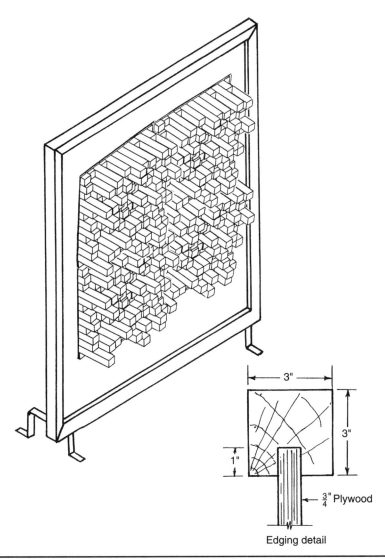

3"

3"

1"

¾" Plywood

Edging detail

FIGURE 12-9   A nontraditional baffle using a diffusing surface.

from the 4 × 4-ft diffusing area is different from that returned from a flat reflective surface in the following ways: the return is about 8 dB lower in intensity; the return is greatly diffused through the half-space; and the return is spread over several milliseconds of time. Musicians may sense a fullness to the sound. To achieve this, the baffle in Fig. 12-9 has four Skyline diffusing modules (manufactured by RPG Diffusor Systems, Inc.) mounted on one side. These particular modules are primitive-root diffusers; other kinds of diffusers, commercial or otherwise, could be employed. Or, the surface could be left as a flat reflector.

CHAPTER 13

Recording Studio for Classical Music

A recording studio for classical music must be large enough to accommodate many musicians playing simultaneously. A recording studio large enough to seat a full symphony orchestra and provide the acoustical space to do justice to such music is beyond the scope of this book. More modestly, we will consider a classical music studio large enough to provide a proper acoustical environment for the recording of chamber music groups, small orchestras, and choral groups. Such a studio could also accommodate a modest-sized symphonic band; however, bands in general need brighter acoustics than symphonic ensembles. Also, with some compromises, such a space could be designed for multipurpose use; for example, it could be used for both recording and performance. The latter could be achieved if temporary seating is supplied.

Design Criteria

Clearly, a studio designed for classical music is not limited to that particular kind of music. Rather, it refers to a studio with sufficient volume to record live acoustical music performed by many musicians. A studio designed for recording classical music is generally larger than a pop music studio, and has a longer reverberation time. It must be large enough to avoid the cramped conditions medium to large music groups encounter when forced to record in many recording studios. In particular, this recording studio for classical music will have a volume of 100,000 ft³. In the design presented here, no specific audience area is provided, although some listeners could be seated informally.

Reverberation Time

Reverberation time of music halls and auditoria has changed from being considered the most important criterion of quality to being considered as one of numerous criteria. The importance of reverberation time has been diminished, but it is still one of the factors to be studied in the design of music (and other) spaces.

As we have noted, reverberation time is that time required for sound in a space to decay 60 dB. If surfaces are highly reflective, this time will be several seconds. If the surfaces are very absorptive, reverberation time will be a fraction of a second. Reverberation time, and the sound quality of the reverberation itself, is more important in classical music recording than in any other music recording. The sound quality of the natural reverberation of the space is integral to the sound of the performing ensemble. In contrast, for example in pop music recording, the studio's natural reverberation is intentionally suppressed so that artificial reverberation can be creatively applied. Because of the short reverberation time of a pop studio, the quality of its reverberation is similarly less important than in a classical music studio.

The reverberation times that best serve the various types of classical music and speech are shown in Fig. 13-1. These plots represent the opinions of many music experts, but must be approached with some skepticism, because the experts are not in full agreement. The concert studio plot is close to the type of space of interest for a recording studio for classical music. Organ music requires a longer reverberation time, while chamber music requires more intimate acoustics associated with a shorter reverberation time. Speech fares best in spaces with a short reverberation time so that speech intelligibility is preserved.

For a recording studio for classical music of 100,000-ft³ volume, a reverberation time of about 1.6 sec is indicated. This will be taken as the target value for preliminary calculations in this design.

Diffusion in the Recording Studio

For hundreds of years, diffusion of sound has been considered to be of importance in music halls. It has traditionally been achieved by ornate plaster work, statuary, and geometric irregularities in the side walls and ceilings of the music spaces. This design

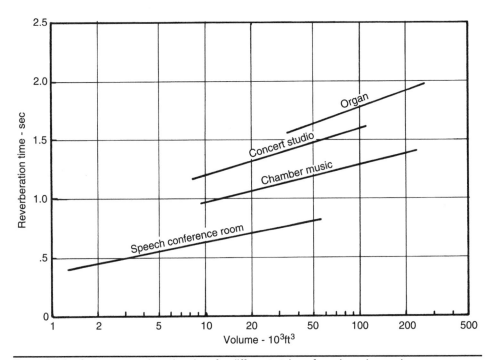

Figure 13-1 Optimum reverberation time for different styles of music and speech.

concept was changed by the availability of Schroeder's reflection phase-grating diffuser, which is far more efficient than geometrical protrusions. In this recording studio for classical music, diffusing elements are placed on three walls and the ceiling. Diffusion is considered in more detail in Chap. 4.

Studio Design

A studio design for recording classical music is shown in Fig. 13-2. The dimensions of the room are $70.0 \times 48.0 \times 30.0$ ft; the room volume is thus 100,800 ft^3. The six surfaces of the room are shown in Fig. 13-2 with all major acoustical treatment. Some suggested treatments use proprietary modules and materials; while these will certainly achieve the design goals, more common construction and materials could be substituted.

The following treatments and features are used in the classical music studio shown in Fig. 13-2: (1) RPG Abffusor, 4×4 ft, quantity 52. (2) NDC Almute, 4×4 ft, with 2 in $\times$ 4 ft $\times$ 4 ft Owens-Corning 703. (3) Tectum Designer, 1-1/2 in $\times$ 24 in $\times$ 48 in, E-405 Mounting, quantity 100. (4–7) Tectum Fabri-Tough wall panels, C-40 Mounting. (4) 12 ft $\times$ 28 ft.

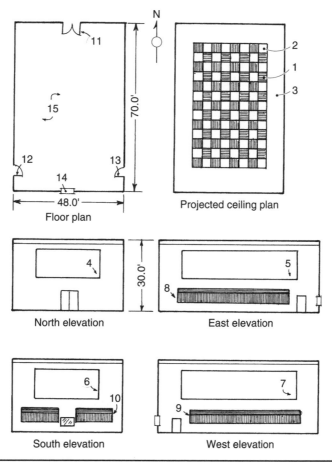

Figure 13-2   Plan and elevation views for a classical music recording studio.

(5) 12 ft × 50 ft. (6) 12 ft × 28 ft. (7) 12 ft × 50 ft. (8–9) RPG QRD-734, 2 ft × 4 ft, quantity 36, rack mounting (see Fig. 13-3 below). (10) RPG QRD-734, 2 ft × 4 ft, quantity 24, rack mounting (see Fig. 13-3). (11) Overly swinging door pair, 2-1/2-in thick, STC-50. (12–13) Overly single swinging door, 1-3/4-in thick, STC-50. (14) Control room observation window, 4 ft × 6 ft. (15) Floor, wood parquet on concrete.

This plan is considered as a stand-alone building; an external control room could be added on the south end of the structure. As shown in the plan, an observation window could be located there. The design of control rooms is considered in Chap. 15.

The building comprises a concrete-block structure with double-leaf walls for insulation from outside noise, and pillars to support the roof structure. The pillars are flush on the inner walls. The specification of the walls is to await an evaluation of the exterior noise exposure, but a Sound Transmission Class of STC-50 for the walls is considered a minimum. Sound isolation is particularly important in classical music recording because of the soft sound levels that are often present; high insulation is mandated. Sound isolation is considered in more detail in Chap. 6. In addition, the heating, ventilating, and air-conditioning (HVAC) system for the building must be designed to provide low background noise; HVAC is discussed in Chap. 10.

A large door on the north end is wide enough for a light truck to enter for transport of instruments and equipment. It is closed by a double swinging door of STC-50 rating to match the walls. Single swinging doors on the south end of both the east and west walls are also rated at STC-50.

Although studio wiring and lighting is beyond the scope of this book, it is worth noting that fluorescent-tube lighting should be avoided in a studio. As noted, particularly for classical-music recording, ambient noise levels must be extremely low. The ballast reactors used in fluorescent lights can generate both an acoustical buzz and electrical interference. If fluorescent lights are necessary, the ballasts should be isolated in metal boxes or placed outside the studio. Many studios have fluorescent-tube lights as bright

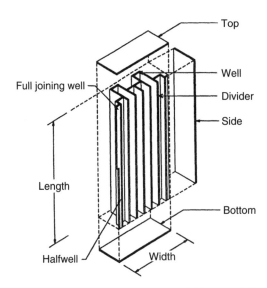

FIGURE 13-3 QRD-734 diffuser construction.

work lights, but they are turned off during recording sessions and other lighting such as track lighting is used instead. Light dimmers must be selected with care; some active-circuit types generate electrical noise; passive resistive rheostats do not.

Acoustical Treatment

This recording studio design uses several proprietary acoustical treatments such as diffusing surfaces on the walls and in the suspended ceiling. The wall diffusing surfaces are built around the QRD-734 diffuser (manufactured by RPG Diffusor Systems, Inc.). This diffuser is described in Chap. 4, and an exploded view of a single unit is shown in Fig. 13-3. This basic quadratic-residue diffuser of prime 7 is fitted with top and bottom plates, and with end plates as required.

A basic cluster of three such 2 × 4-ft units is shown in Fig. 13-4. The top horizontal unit diffuses sound in the vertical direction, and the lower units with vertical wells diffuse sound in the horizontal direction. Between the two, full two-dimensional diffusion is accomplished.

The racks of QRD-734 modules on the walls are supported on 3/4 × 3-in wall cleats as shown in Fig. 13-5. A 48-ft row of QRD-734 modules is placed on the east wall and the west wall. A 16-ft row is placed on either side of the control room observation window on the south wall. A total of 96 individual 2 × 4-ft QRD 734 modules are used.

The placement of the rows of QRD-734 modules on the walls is critical to prevent flutter echoes between opposing, parallel, reflecting surfaces. Only horizontal rays of sound can excite flutter echoes. If all sound sources (such as musical instruments) are 2 to 4 ft above the floor, the bottom of the diffusers should be about that distance from the floor. In Fig. 13-2, modules (8), (9), and (10) are shown 4 ft above the floor. It is possible that these should be lowered, but typically there are so many scattering sources close to floor level between the two reflecting surfaces that flutter echoes that low are unlikely. Although proprietary diffuser modules are shown as examples in this design, other types of diffusion could be employed or similar quadratic-residue diffusers could be constructed from scratch using the equations given in Chap. 4.

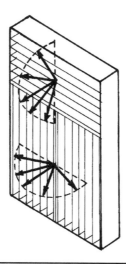

FIGURE 13-4   Diffusion of three-group QRD-734 diffuser modules.

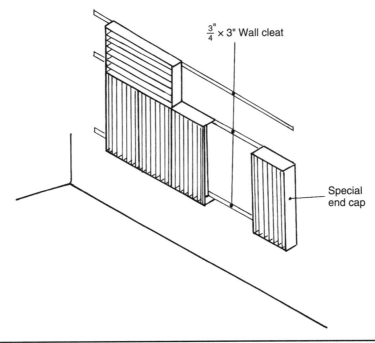

$\frac{3"}{4} \times 3"$ Wall cleat

Special
end cap

FIGURE 13-5   Wall mounting of QRD-734 diffuser modules.

Wall Panel Absorbers

For general absorption, wall panels (4), (5), (6), and (7) of Fig. 13-2 have been included in the design. These panels could be made of any one of a dozen brands of commercial absorbing panels, but Tectum Fabri-Tough panels (manufactured by Tectum Inc.,) of 1-in thickness have been selected for this design. These panels are installed with standard C-40 mounting, that is, with an airspace of 40 mm (1-1/2 in) between the panels and the wall. This increases the low-frequency absorption materially. The absorption coefficient α of these panels will be about 0.30 at 125 Hz, 0.77 at 250 Hz, and essentially 1.0 through 4 kHz. If measurements and experience show that the reverberation time of the studio should be lowered or increased, it is easy to increase or decrease wall panel area. Alternatively, low-cost porous absorber panels can be constructed from scratch, as described in Chap. 3.

Ceiling Treatment

The suspended ceiling uses the standard C-400 mount, in which the ceiling is dropped down 400 mm (16 in). This ceiling is of the type illustrated in Fig. 13-6. This kind of mounting serves multiple functions. In addition to allowing space for heating, ventilating, and air-conditioning (HVAC) ducting, lighting units could be accommodated. However, in this studio drop lighting units are considered more desirable as they bring the light sources closer to the performers and their music stands.

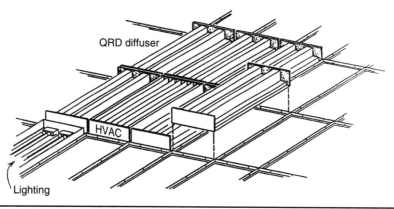

Figure 13-6 T-frame mounting of modules.

In this design, both diffusing/absorbing and ordinary absorbing units are mounted in the suspended ceiling frame. For example, the RPG Abffusor (manufactured by RPG Diffusor Systems, Inc.) is both a diffuser and an absorber. The Abffusor combines the high-and mid-frequency properties of porous materials with the low-frequency properties of diaphragmatic membranes to give good absorption down to 100 Hz for all angles of incidence. The absorption coefficients of the Abffusor and other materials used are listed in Table 13-1.

There are three types of lay-in panels in the suspended T-frame. One, as described above, is the Abffusor. The second is an NDC Almute panel (manufactured by Peer Inc.) with 2-in of 6 lb/ft³ glass fiber (Owens Corning 705) on it. The Almute panel is a 2.5-mm (3/32-in) panel fabricated from sintered aluminum particles with 100-micron holes. The glass-fiber backing should not touch the panel. The 52 Abffusors and 52 Almute panels fill the central portion of the suspended ceiling.

The third type of lay-in panel surrounds the central portion of the suspended ceiling. These are 424 panels of 2 ft × 2 ft × 1-1/2 in Tectum Designer (manufactured by Tectum Inc.) panels. This 8- to 10-ft band of Tectum around the edge of the ceiling serves to place the diffusing elements nearer the center of the room for greater effect.

Reverberation-Time Calculations

As with any room, the reverberation time of this recording studio should be verified by measurements. Calculations can give only an approximation of reverberation time because of the many factors that can enter between theory and practice. Table 13-1 lists the principal contributions to the absorption of the room. The table is divided into the suspended ceiling absorption and the wall absorption.

The absorption coefficient α of each material is multiplied by the surface area S of the material to obtain the $S\alpha$ product for each material at six standard frequencies. The $S\alpha$ product yields the number of absorption units (sabins) contributed by each material area. The sum of the absorption units is added to find the total absorption units at each frequency. The Sabine equation for reverberation time is:

$$RT_{60} = 0.049 V / A \qquad (13\text{-}1)$$

Table 13-1 Reverberation-Time Calculations for a Classical Music Studio

Description	Area, ft²	125 Hz		250 Hz		500 Hz		1 kHz		2 kHz		4 kHz	
		α	Sα	α	Sα	α	Sα	α	Sα	α	Sα	α	Sα
Suspended ceiling:													
RPG Abffusor	208	0.82	171	0.90	187	1.00	208	1.00	208	1.00	208	1.00	208
NDC Almute with 2-in 703	208	0.90	187	0.92	191	0.99	206	1.00	208	1.00	208	1.00	208
Tectum Designer	1696	0.44	746	0.47	797	0.36	671	0.51	865	0.71	1209	1.00	1696
Wall treatment:													
RPG QRD-734	768	0.23	177	0.24	184	0.35	269	0.23	177	0.20	154	0.20	154
Tectum Fabri-Tough C-40	1872	0.30	562	0.77	1441	1.00	1872	0.98	1835	0.79	1479	0.95	1778
Concrete block, paint	4288	0.10	429	0.05	214	0.06	257	0.07	300	0.09	386	0.08	343
Floor, parquet on concrete	3360	0.02	67	0.03	101	0.03	101	0.03	101	0.03	101	0.02	67
Total absorption, sabins			2339		3115		3584		3694		3740		4454
Reverberation time, sec			2.11		1.59		1.38		1.34		1.32		1.11
Air absorption (50% RH)									—	3.0/k	302	8.0/k	806
Total absorption (with air)			—		—		—		—		4042		5260
Reverberation time, sec (50% RH)			2.11		1.59		1.38		1.34		1.22		0.94

where RT_{60} = reverberation time, sec
  V = volume of room, ft³
  $A = S\alpha$ = total absorption of room, sabins

For this particular room, with volume of 100,800 ft³, Eq. 13-1 yields:

$$RT_{60} = (0.049)(100,800)/A$$
$$= 4939/A$$

The concrete block walls are painted, resulting in a minor (but not to be neglected) decrease in sound absorption.

Table 13-1 also shows the calculations for the reverberation time of the recording studio. The time is first calculated without air absorption, which is added later. The plotted results of these calculations are shown in Fig. 13-7. Although RT_{60} is calculated at six frequencies, single-figure appraisals of reverberation time are commonly made at 500 Hz. The 500-Hz reverberation time in this case is about 1.4 sec (without air absorption). Figure 13-1 shows that a concert studio of 100,000 ft³ size calls for a reverberation time of about 1.6 sec. Chamber music, on the other hand, calls for a reverberation time of about 1.3 sec. With the uncertainty of different types of music as well as the uncertainty of calculations versus measurements, the preliminary value of 1.4 sec is reassuring.

Air Absorption

Because of air absorption, sound is attenuated as it travels through air. The attenuation is significant in large rooms where sound path lengths are long. Air absorption should be considered in this large studio. Its effect is only at higher frequencies, but

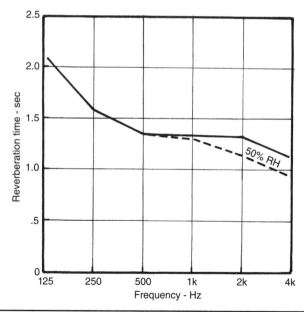

Figure 13-7   Reverberation time of a classical music studio.

the resulting decrease in reverberation time at higher frequencies should be documented. In Table 13-1, air absorption is added at 2 kHz and 4 kHz. Air absorption can be approximated from:

$$A_{air} = mV \tag{13-2}$$

where A_{air} = air absorption, sabins
  m = air attenuation coefficient, sabins/ft³
  V = volume of room, ft³

The value of the air attenuation coefficient m varies with humidity. With humidity between 40% and 60%, the values of m at 2, 4, and 8 kHz are: 0.003, 0.008, and 0.025 sabins/ft³.

It is also important to note that people are absorptive. Some acousticians use a rule of thumb that each seated person adds 5 sabins at 500 Hz. In a recording studio such as this, the number of musicians will not greatly affect absorption and reverberation time. However, in a performance hall, with many hundreds of people, audience absorption will play a significant role; concert halls sound quite different depending on whether they are empty or full.

Initial Time-Delay Gap

The initial time-delay gap (ITDG) is the short span of time between the arrival of the direct sound and the reflected components arriving shortly thereafter. This gap is imprinted on every music signal and has an important effect on the sound quality of the music. This is true both for live music performance (musician and listener), and in music recording (musician and microphone). Experts in the fields of music and concert-hall acoustics tend to agree that the initial time-delay gap is related to the subjective impression of the intimacy of the music. The term "initial time-delay gap" was coined by Beranek in the 1960s; in some cases, the term "arrival-time gap" is used.

In a concert hall, a person seated in the audience first hears the onset of the direct sound, followed by reflections from the side walls, ceiling, and so on. The most revered concert halls generally have initial time-delay gaps in the range of 20–30 msec. Those reflections in the concert hall arriving laterally have been demonstrated to be related to the impression of spaciousness of the music.

Figure 13-8 explores the issue of initial time-delay gap in a studio; the figure is based on the dimensions of the example music recording studio. Taking a spot in the orchestra as the location of the source, the direct sound travels a 10-ft path to the microphone. Reflections from the ceiling travel about 41 ft to reach the microphone. Taking a time of 0 sec as the time the sound leaves a musician in the orchestra, the direct 10-ft path takes 10 ft/(1130 ft/sec) = 9 msec for the sound to reach the microphone. The reflection takes 41/1130 = 36 msec to reach the microphone. The arrival times of the direct and reflected components may then be plotted on the graph of Fig. 13-8 at 9 and 36 msec, respectively. The side wall reflections may be plotted beginning at 43 msec. The arrival of the direct ray opens the gap and the arrival of the first strong reflections closes the gap. The gap will not be silent, but will be filled with a low-level clutter. As long as this clutter is 20 to 30 dB below the peaks, the gap will serve its intended purpose of creating the impression of intimacy in the sound field.

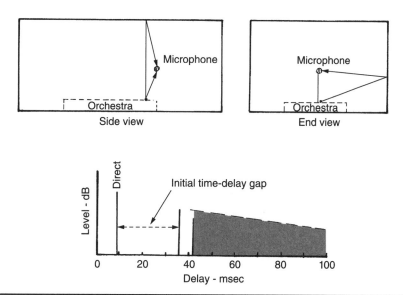

FIGURE 13-8   Initial time-delay gap in a classical music studio.

Comments on Design

In this example, although the calculated reverberation time meets the optimum design goal, reverberation time is a slender reed on which to lean. This design actually depends on reverberation time only to arrive at a reasonable distribution of absorbing and diffusing elements. On the other hand, this design does not particularly depend on diffusion. Generally diffuse conditions should result from the ceiling and the walls, and this is expected to be the main characterization of this studio.

What is the effect of the diffusion designed into this studio? The individual in a well-diffused space has the impression of being in a larger space. Musicians should enjoy excellent ensemble, that is, the individual musician should be able to hear the other musicians, which is necessary for coordination. Music in a diffuse space has an open, free, uncluttered character that forms the basis of a good recording. Of course, microphone selection and placement, as well as the entire art and skill of classical music recording, will determine the final success of the recording.

CHAPTER 14

Voice Studio

Some studios are specifically designed for recording spoken voice. In particular, voice-over recording is that process by which a narration is recorded to be used alone, or mixed with background music and/or sound effects. It can also be applied to the simple recording of voice alone. The studio for doing this is called a voice studio, and it has its own particular acoustical characteristics, just as a music studio has its own acoustical requirements. Similarly, automatic (or automated) dialogue replacement (called ADR) is a process by which dialogue is recorded in sync with a movie's visuals. Because of background noise, original voice recordings made on a shooting location are often unusable; these lines of dialogue are re-recorded in small voice studios. In animated films, the dialogue is recorded first, and the animation is created to fit the dialogue.

At first glance it might seem that the acoustical requirements for a voice studio would be simpler than those of a music studio. This is not necessarily true. Voice sounds are subject to colorations (such as those caused by modal resonances and comb-filter effects), which can also affect music, but which are usually less audible in music because of its complexity. Although a relatively dead ambience is preferred, a voice studio should not be completely dead. This is especially true because in some cases voice studios are used to record singing. As with any studio, flutter echo must be eliminated by splaying walls, applying diffusion, or by other means. In other words, a voice studio presents its own set of acoustical challenges.

Design Criteria: Isolation

As with any recording studio, a voice studio must be acoustically isolated. Because voices operate at low volume, it is quite easy to prevent voices from interfering with activities outside the studio. For example, several voice studios could be built adjacent to one another and not suffer from crosstalk problems. Simple gypsum-board stud walls would probably suffice.

Simple gypsum-board walls may also block low-volume outside noise from entering the studio and interfering in the speech region. However, these walls may not block loud noises, or noises lower in frequency. Thus, if this is the case, heavy masonry and more complex walls will be needed. More economically, if the interfering noise is particularly outside the speech frequency region (for example, low-frequency structure-borne noise from an adjacent HVAC room), electrical filters could be placed in the speech recording signal path to block noise from the speech recording. However, the acoustical noise would still be audible to those working in the voice studio.

Design Criteria: Room Size

Most voice studios are very small with modest floor space. Although economic factors can outweigh the acoustical, there is a penalty in having a too-small studio. The modal resonances of the room dictate the acoustics of the room, as described in Chap. 5. If the room is too small, modal resonance frequencies will be too few, with too-great spacing between them. This becomes a permanent flaw of the room, with no satisfactory correction. The smaller the room, the higher the low-frequency limit; that is, the lowest frequency with resonance support. For example, a room 10-ft long will have a bass limit of 56 Hz, but a 20-foot room will have a bass limit of 28 Hz.

Above 300 Hz the modal frequencies are so close together that problems associated with the lower frequencies tend to disappear. Acoustical engineers of the British Broadcasting Corporation, on the basis of voice coloration studies of their hundreds of studios, determined that it is impractical to build voice studios with volume less than 1500 ft^3.

Design Criteria: Room Shape

In this rectilinear world, a rectangular shape is assumed, but what dimensional proportions should be selected? Scores of papers have been written presenting arguments about why certain room proportions give the most uniform distribution of room modes. All have strong and weak points and none result in the perfect distribution of modal frequencies. Here are three proportions that have stood the test of time:

	Height	Width	Length
(A)	1.00	1.14	1.39
(B)	1.00	1.28	1.54
(C)	1.00	1.60	2.33

Assuming a ceiling height of 10 ft, these three proportions offer the following:

	Height	Width	Length	Volume
(A)	10.0 ft	11.4 ft	13.9 ft	1585 ft^3
(B)	10.0	12.8	15.4	1971
(C)	10.0	16.0	23.3	3728

Proportion A gives a volume that has been classed as marginal in size for a room to be used for a voice studio. Proportion B is a bit better, but proportion C ($10 \times 16 \times 23.3$ ft) is selected as the most promising and will be used in this chapter.

Axial-Mode Study of Selected Room

As we observed in Chap. 5, of the three types of room modes, axial modes contain the most energy, and thus are the most significant. An appraisal of the axial modes alone can give a quick judgment of the acoustical quality of the room. Table 14-1 lists the axial modes of the $10 \times 16 \times 23.3$-ft room.

The lowest axial mode can be found from $f = 1130/2L$, in which 1130 is the speed of sound in ft/sec and L is the length of the room, 23.3 ft: $f = 1130/(2 \times 23.3) = 24.2$ Hz. Integral multiples of 24.2 Hz extend up through the spectrum, but we stop at 300 Hz

	Length $L = 23.3$ ft $f_1 = 565/L$	Width $W = 16$ ft $f_1 = 565/W$	Height $H = 10$ ft $f_1 = 565/H$	Arranged in Ascending Order	Difference
f_1	24.2 Hz	35.3 Hz	56.5 Hz	24.2 Hz	11.1 Hz
f_2	48.5	70.6	113.0	35.3	13.2
f_3	72.7	105.9	169.5	48.5	8.0
f_4	97.0	141.3	226.0	56.5	14.1
f_5	121.2	176.6	282.5	70.6	2.1
f_6	145.5	211.9	339.0	72.7	24.3
f_7	169.7	247.2		97.0	8.9
f_8	194.0	282.5		105.9	7.1
f_9	218.2	317.8		113.0	8.2
f_{10}	242.5			121.2	20.1
f_{11}	266.7			141.3	4.2
f_{12}	291.0			145.5	24.0
f_{13}	315.2			169.5	0.2
				169.7	6.9
				176.6	17.4
				194.0	17.9
				211.9	6.3
				218.2	7.8
				226.0	16.5
				242.5	4.7
				247.2	19.5
				266.7	15.8
				282.5	0.0
				282.5	8.5
				291.0	24.2
				315.2	
				Mean difference: 11.64 Standard deviation: 7.1	

TABLE 14-1 Axial Modes in a Voice Recording Studio (Room Dimensions: 23.3 ft × 16 ft × 10 ft)

because modes become densely spaced; therefore, modal problems are rare above that frequency. These are the characteristic frequencies associated with the standing waves between the front and the rear walls. Similar computations are made for the 16-ft width and the 10-ft height.

The most important consideration about axial modes is their spacing. For this reason, to study this spacing, all three sets of axial modes are arranged in ascending order. The right column of Table 14-1 isolates mode spacings for evaluation. The mean spacing is 11.6 Hz, which looks good when compared to 25 Hz, above which modal distortions will likely be audible.

There is a coincidence (two axial modes) at 282.5 Hz, which probably will not be audible because it is so close to the arbitrary 300-Hz limit. There is another near coincidence at 169.5 Hz, which can possibly pose a greater potential problem. Otherwise the prospects are excellent, with a warning to be alert to possible coloration of voice timbre around 170 Hz.

Acoustical Treatment

An elevation sketch of the proposed $10 \times 16 \times 23.3$-ft room is shown in Fig. 14-1. Clearly, the human speaker will be close to the microphone. Most of the sound at the microphone will be direct sound from the voice. However, sound will also reflect from the room's surface areas and be received at the microphone with varying degrees of delay. Thus, the surface treatments must be carefully considered.

First, diffusing elements will be placed at the right end of the room, as shown in the figure. There are two options for treating the end of the room near the microphone. In either case, steps must be taken to control early reflections, which are a potential problem. Briefly, these reflections can be absorbed collectively (soft option), or treated on an individual basis (hard option). These two options are discussed below.

Early Reflections

The voice of the person at the microphone in Fig. 14-2 is reflected from nearby surfaces. Each first-order reflection impinges on a specific area of floor, side or end walls, and ceiling. Corner reflectors formed by the intersection of two or three surfaces return their energy directly back to the source.

The first reflection to return to the microphone is reflection (1) from the floor. It can be of significant magnitude because of the short path length. The ceiling reflection (2) is directly over the source. The three vertical walls around the source send three more reflections (3), (4), and (5) back to the microphone. These five returns have undergone a single, normal (right-angle) reflection.

The eight corner reflections (6) through (13) are from the intersection of two planes, while the four reflections (14) through (17) are from triplanes. Those from two-plane intersections undergo two reflections; those from triplane intersections undergo three reflections. As there is loss at each reflection, the corner reflections are of lower relative

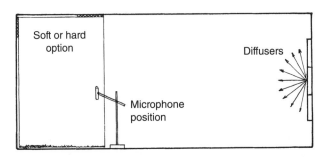

FIGURE 14-1   Elevation view of the voice recording studio.

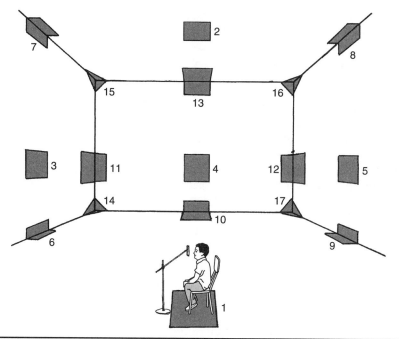

FIGURE 14-2 First-order reflections in the voice studio.

level than the normal reflections, and the triplane reflections are lower than the two-plane reflections.

Reflections (1) through (5) are dominant early first-order reflections. They are the earliest in time and the greatest in magnitude. The corner reflections are later and lower, but cannot be neglected in some circumstances.

The Soft Option and Hard Option

The direct voice signal is picked up at the microphone as shown in Fig. 14-1. After the direct voice signal arrives, reflections (1) through (5) arrive at the microphone with good amplitude, and then (6) through (17) with reduced amplitude. After that, multiple reflections (not shown in the figure) arrive from nearby walls and, later, from distant walls (including the far end of the room). Research has revealed the effect of the early lateral reflections on the direct signal (see Fig. 11-16). In particular, when lateral reflections are correctly controlled during the first 60 msec, the sound can achieve a natural sense of spaciousness.

The soft (less reflective) option plan is to cover all surfaces of the entire left end of the room (Fig. 14-1) with sound-absorbing material. This would reduce the amplitude of all early reflections. However, it can also increase the overall absorbence of the room too much, and make the room too dead.

The hard (more reflective) option plan would leave the entire left end of the room (Fig. 14-1) with untreated surfaces except for the specific spots where reflections strike the walls, floor, or ceiling and then arrive at the microphone. A rug is placed under the microphone to reduce the floor reflection (1). Reflections (2) through (5) are reduced by placing a 2 × 2-ft absorber at spots where they strike room surfaces. This will be sufficient to avoid artifacts in the signal. Because of their lower amplitude, it probably will not be necessary to use absorbers to specifically address reflections (6) through (17), but their presence is acknowledged. Reducing the amplitude of reflections (1) through (5) has been very successful in listening rooms. [Referring again to Fig. 11-16, there remains the question of whether the early reflections should be reduced below the audibility threshold curve (A) or to the intermediate curve (B), which would give a sense of spaciousness to the signal.]

The soft option would tend to reduce all early signals below the threshold of audibility. This is a tremendous improvement over having all reflections mixed with the direct signal, but is it the best? Subjective evaluation in the (nearly completed) room would need to be performed to determine more precisely the extent to which early sound should be reduced.

Diffusion

Diffusion in the studio (Fig. 14-1) can be obtained in a number of ways. For example, Skyline diffusing modules (manufactured by RPG Diffusor Systems, Inc.) can be used. Each 24 × 24-in unit is made up of a 12 × 13 matrix of small blocks of thermoformed polymer to form a two-dimensional diffusing surface based on a primitive-root, number-theory sequence (see Chap. 4). The right end of the studio (Fig. 14-1) supports a three-high by four-wide array of these 2 × 2-ft omnidirectional units.

These diffusing surfaces will form a solid basis for recording high quality voice signals in this studio once the design of the microphone end of the room is decided. That determination requires a bit more consideration of general studio characteristics.

Mode Control

The best approach for controlling the low modal frequencies is to apply absorbing materials or resonant structures. The optimal location of low-frequency absorption in a room is not always obvious; corner placement is often effective, but experiments may show that wall placement is better. In this suggested design, two absorbers are mounted on the far wall adjacent to the Skyline diffusing surfaces, one on each side, as described below. For example, Modex Plate absorbers (manufactured by RPG Diffusor Systems, Inc.) can be used. These membrane-type absorbers offer attenuation in the low-frequency region required for normal mode control; in this example, Type A mounting is used. The absorption characteristic of a Modex Plate is shown in Fig. 14-3.

Studio Design: Soft Option

Figure 14-4 is an "exploded" view of the voice studio, showing the floor plan and the four walls laid out. This illustrates the soft (less reflective) option for the room. The structural construction of the room is assumed to be gypsum board on a wood frame, with a wood floor. It is further assumed that the walls have adequate STC rating (see Chap. 7) to provide an ambient noise level within the studio of about NCB-15 (see Chap. 11). This will require close attention to the heating, ventilating, and air-conditioning (HVAC)

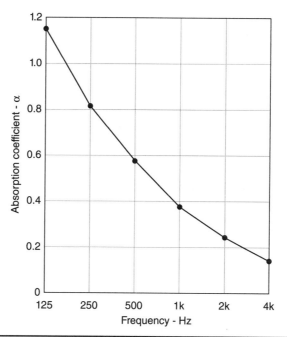

FIGURE 14-3   Absorption characteristic of a Modex Plate absorber.

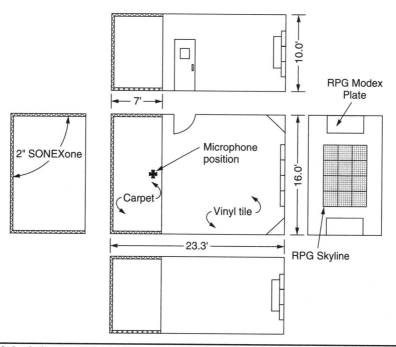

FIGURE 14-4   Soft-option voice studio design.

equipment mounting and duct planning (see Chap. 10). Other features such as the Modex Plate absorbers and Skyline diffusers are present in the room.

Figure 14-4 features an absorptive end of the room where the microphone will be placed. The floor at this end is covered with heavy carpet and pad out to the 7-ft point. There are many types of sound absorption materials suitable for the soft-end walls and ceiling. One possibility is 2-in SONEXone (manufactured by Pinta Acoustic, Inc.) (see Chap. 3). It should be remembered that the voice reflections are of mid to high frequency, and most sound-absorbing materials, including porous absorbers, are highly effective in this upper range.

Studio Design: Hard Option

Figure 14-5 specifies conditions for the hard option (more reflective) for the voice studio. As before, two Modex Plate absorbers are mounted on the wall far from the microphone. The 12 Skyline diffusing modules are also mounted on the same end. An area rug (1) is placed under the microphone to conform to the absorber location in Fig. 14-2. The ceiling absorber (2) is mounted directly over the microphone position. Then come the side wall and end wall absorbers (3), (4), and (5). All of these absorbers measure approximately 2 × 2 ft, and (3), (4), and (5) are centered at the microphone level, approximately 4 ft above the floor. This completes the specific treatment for the hard-option voice studio. Note that there is very little general absorption in the room. The question must be considered: Will it be too live?

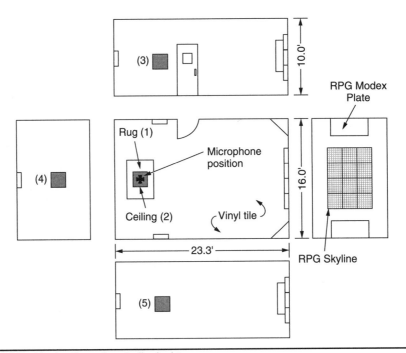

Figure 14-5 Hard-option voice studio design.

Reverberation-Time Calculations

Reverberation-time (RT_{60}) calculations will help determine the suitability of the two absorption options. The reverberation times of the soft option and the hard option are calculated separately using the Sabine equation. The studio volume is 3728 ft³. The complete calculations are shown in Table 14-2.

Section A of Table 14-2 covers the natural absorption of the room, which can easily be overlooked. The gypsum-board walls and the wood floor act as diaphragms, and when vibrating they absorb sound. The coefficients for gypsum-board walls and wood floor are found in tables in the Appendix. Separate calculations are made at each of the standard six frequencies for which coefficients are available. It is noted that the absorption in the room due to the bare walls and floor varies from 159.9 sabins at 125 Hz to 57.3 sabins at 4 kHz.

Section B of Table 14-2 applies to the surface treatments at the microphone end, which give it the absorbent characteristics. The absorption of SONEXone and carpet varies from 70.8 sabins at 125 Hz to 441.6 sabins at 4 kHz.

Section C of Table 14-2 applies to the Modex Plate absorbers. The absorption coefficients are high particularly at low frequencies, but the area is small so the total absorption is relatively low. Clearly, if needed, adding additional units would increase low-frequency absorption. Section D of Table 14-2 shows the absorption of the spot absorbers (1) through (5) mounted around the microphone to reduce amplitude of first-order reflections.

Reverberation Time: Soft Option

The room absorption that applies to the soft option in Table 14-2 is listed as the sum of sections A, B, and C (i.e., the natural absorption of the walls and floor, the SONEXone, the carpet, and the Modex Plates). The total absorption in sabins for these three sections is given in Table 14-2 for all six frequencies. The calculated reverberation time is plotted as the lower curve in Fig. 14-6. The curve flattens off at a reverberation time of about 0.35 sec at 1 kHz, which should be appropriate for this space. There seems little chance that anyone would desire a reverberation time less than 0.35 sec. There is no simple way of increasing the reverberation time without decreasing the absorbing material around the microphone, and that is exactly what has been done in the hard option that follows.

Reverberation Time: Hard Option

The room absorption in Table 14-2 that applies to the hard option gives the total absorption as the sum of A, C, and D with only the five 2 × 2-ft panels of 2-in absorbing material on the reflection spots [in practice, the rug spot (1) should be a bit larger]. The absorption totals in sabins for each frequency are used to calculate the reverberation time. These reverberation times are plotted on the graph of Fig. 14-6 as the upper broken curve.

Reverberation time for this hard option ranges from 0.93 sec at 125 Hz to 2.26 sec at 2 kHz. Few recording engineers will be happy with this too-bright condition, and it is

	Description	Area, ft²	125 Hz α	125 Hz Sα	250 Hz α	250 Hz Sα	500 Hz α	500 Hz Sα	1 kHz α	1 kHz Sα	2 kHz α	2 kHz Sα	4 kHz α	4 kHz Sα
A	Walls and Ceiling	1039	0.10	103.9	0.08	83.1	0.05	52.0	0.03	31.2	0.03	31.2	0.03	31.2
	Floor, wood	373	0.15	56.0	0.11	41.0	0.10	37.3	0.07	26.1	0.06	22.4	0.07	26.1
	Absorption, sabins, untreated			159.9		124.1		89.0		57.3		53.6		57.3
	Reverberation time, sec, untreated			1.15		1.47		2.05		3.19		3.41		3.19
B	2-in SONEXone	412	0.15	61.8	0.34	140.0	0.81	333.7	1.00	412.0	0.92	379.0	0.90	371.0
	Carpet, heavy	112	0.08	9.0	0.27	30.2	0.39	43.7	0.34	38.1	0.48	53.8	0.63	70.6
	SONEXone + carpet			70.8		170.2		377.4		450.1		432.8		441.6
C	Modex Plate	32	1.15	36.8	0.83	26.6	0.58	18.6	0.38	12.2	0.25	8.0	0.16	5.1
D	Spot absorbers	20	0.02	0.4	0.22	4.4	0.69	13.8	0.90	18.0	0.96	19.2	1.0	20.0
	Soft-Option Design (A + B + C)			267.5		320.9		485.0		519.6		494.4		504.0
	Reverberation time, sec			0.68		0.57		0.38		0.35		0.37		0.36
	Hard-Option Design (A + C + D)			197.1		155.1		121.4		87.5		80.8		82.4
	Reverberation time, sec			0.93		1.18		1.50		2.09		2.26		2.22

TABLE 14-2 Reverberation-Time Calculations for a Voice Recording Studio

234

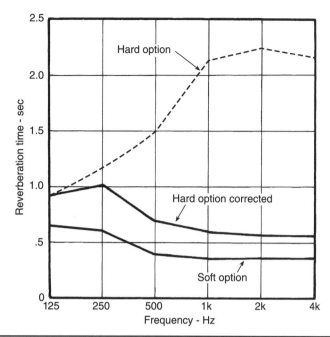

FIGURE 14-6   Reverberation times of soft- and hard-option designs.

assumed that it will be altered. To achieve a reasonable reverberation time of about 0.5 sec, additional absorption is required.

The rising reverberation curve of Fig. 14-6 for the hard option can be corrected after the need is verified by reverberation measurements and listening tests on voice signals recorded in the room. The room will undoubtedly be judged as too reverberant. One possible approach for correcting it is to mount 86 Tectum sound blocks (manufactured by Tectum Inc.), 43 to each side wall. Each block measures 15-1/2 × 15-1/2 × 2 in and is fabric covered. They should be mounted on 24-in centers to achieve the absorption figured into the "hard option corrected" curve of Fig. 14-6; they can be oriented as squares or diamonds.

Sound Field Response

Figure 14-7 shows the time-base sound field response of the hard-option studio before the spot absorbers are installed. The reflections from spot absorbers (1) through (5) are labeled at their respective levels and delays. With the spot absorbers installed, these would be reduced 15 to 20 dB with respect to the direct signal. With the soft-option treatment, these reflections would be reduced about the same amount, 15 to 20 dB. In both options, there is a strong direct signal with close to zero delay. Also in both options, a welcome 30-msec initial time-delay gap prevails before the late reflections and diffusion fall away.

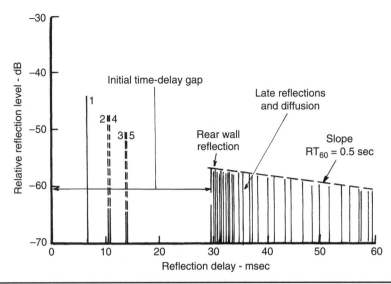

FIGURE 14-7 Sound field response of the hard-option design before spot absorbers are installed.

Comments on Design

This chapter has presented an economical design for a voice studio. Both the soft option of Fig. 14-4 and the hard option of Fig. 14-5 have excellent prospects. The five spot absorbers in the hard-option design reduce the amplitude of the first-order reflections, but do not affect the corner reflections. Do we want to eliminate these reflections, or do we want to adjust their amplitudes to some desirable level? There is room for experimentation in the hard-option design; in the soft-option design, all early reflections are absorbed.

CHAPTER 15

Control Room

The mixing engineer working in a control room is responsible for evaluating and refining the sound from the monitor loudspeakers. The source may be live music from the studio, an existing recording, or the different tracks of a project being mixed. The engineer must judge sound quality using criteria such as frequency response, level changes, distortion, tonal differences, spatial distribution, timing of segments, and scores of other minute details. To accomplish this, the mixing engineer needs a mixing environment that is acoustically neutral. For example, if the control room's acoustics adds a bass boost to the sound, the mixing engineer will instinctively decrease bass in his or her mixes to achieve a balanced result. But when played back in other rooms, the recordings would be bass deficient. To prevent this kind of error, the acoustical link between the monitor loudspeakers and the mixer's ears must minimize any unfavorable perceptual changes. That is the important burden of the control room: to ideally deliver the playback sound without adversely adding anything to it or taking anything away.

In most home playback rooms, much is changed as sound travels from the loudspeakers to the listener's ears. We become accustomed to this form of distortion and may accept it as the "sound" of the room. In a control room, great effort is expended to remove this distortion. Using headphones might be the only way to eliminate all of it, but that is an unsatisfactory and unnatural solution; mixing engineers greatly prefer to work in an open acoustic environment. The control room must be specifically designed to deliver balanced sound at the mixing position.

Early Reflections

Don Davis and Chips Davis first experimented with control room designs in which the front end of the room was completely absorbent (dead), and the back end was diffusive and reflective (live). They noted high quality of sound from the monitor loudspeakers that was both accurate and natural sounding (Davis and Davis, 1980). By applying absorptive glass fiber around the loudspeaker end of the control room, early reflections were reduced along with their degrading effect on sound quality. This room configuration is known as live end-dead end (LEDE) and is used in many control rooms.

Adding absorbent material to control room surfaces is one way to reduce the effect of early reflections, but it absorbs precious sound energy as well. The mixer compensates by increasing the gain to raise the sound level to a desired level. Many control rooms minimize the use of absorbent, with the result that amplifiers and loudspeakers can run at a lower level and with lower distortion. With room designs such as LEDE, absorbent is only placed where it is most needed while other surfaces are reflective; this overcomes problems from reflections but also preserves sound energy.

Combing of Early Reflections

The direct sound arriving at the mixing engineer's ears is clean as far as the acoustical path is concerned (of course, every loudspeaker adds its own distortion to the sound). In the acoustical design, the direct sound is the "best" sound. Sounds reflected from the side walls, ceiling, or floor (or the mixing console itself) arrive with appreciable level differences and at slightly different times. When each of these replica signals with small time differences between them combines with other signals at the mixer's ears, a comb-filter response is created. This is true whether the reflections combine with each other or the reflections combine with the direct sound. The result of this reinforcement and cancellation is to change the nominally flat response to a series of peaks and nulls through the audible spectrum. The frequency spacing of the comb pattern is determined by the time delay between the arrival of the sounds, and the amplitude of the peaks and dips is determined by the relative amplitudes of the sounds.

When measured with single constant frequencies with certain delay times and equally strong reflections, plots of comb filtering make it appear to be very damaging. In practice, in most cases, with complex and transient signals such as music and speech, comb filtering occurring in a room is often not particularly audible. Moreover, as frequency increases, the density of peaks and dips increases and thus combing becomes even less problematic. As noted, in home playback one becomes accustomed to this form of distortion in the signal created by this combing.

However, no matter how minor its effect, this uneven frequency response is undesirable in rooms designed for critical listening. In particular, in professional control rooms, where every audio decision is predicated on the quality of the signal at the mixing engineer's ears, the cause of such distortion must be addressed. Comb filtering can be minimized or prevented by placing an absorber (or diffuser) on the reflecting walls to attenuate or eliminate reflected sounds.

Examples of Comb Filtering

The effect of comb filtering is shown in Fig. 15-1. These plots were taken from a simplified laboratory setup in which a sound source, a microphone, and a baffle (a plywood panel) were arranged in that order in a straight line. In Fig. 15-1(A), the left peak is the

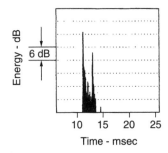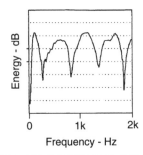

Figure 15-1 An example of comb filtering due to a specular reflection. (A) Time domain. (B) Frequency domain.

direct sound picked up by the microphone on its way to the baffle. The second peak is a specular reflection from the baffle. These two combine with a time (phase) difference between them creating the typical peak/null of a comb-filtered response as shown in Fig. 15-1(B). As noted, this frequency-response distortion is to be avoided in control rooms.

The data plotted in Fig. 15-2 is not from a laboratory setup, but rather an actual condition in a control room. In Fig. 15-2(A), the direct peak on the left and the specular-reflection peak on the right are separated in time by about 7 msec. In combining, the resulting response is changed from a flat condition to a series of peaks and nulls.

In Fig. 15-2(B), the direct spike is the same as the one above, but the reflection (instead of being specularly reflected) has fallen on a diffusing surface, and has been diffused over 12 msec of time. When the diffused reflection combines with the direct spike, some comb filtering results but the amplitudes of peaks and nulls are reduced and the fluctuations are of a higher frequency. From a perceptual standpoint, the response deviations of (A) are audible, but that of (B) are essentially inaudible. Thus, by diffusing early reflections before they combine with the direct sound or other reflections, the audible distortion can be practically eliminated. Alternatively, the classic solution is to place absorption on the pertinent reflective surfaces; by attenuating the reflections, their effect is reduced. However, this method also reduces the overall sound energy in the room.

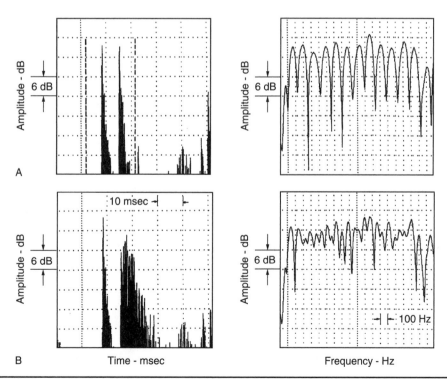

Figure 15-2   The effect of diffusion on comb filtering. (A) Specular reflection in time and frequency domains. (B) Diffused reflection in time and frequency domains.

Reflection-Free Zone

The concept of a reflection-free zone (RFZ) is simple: To eliminate or attenuate early reflections arriving at the mixing position. A reflection-free zone around the mixing position can be achieved by covering the offending reflecting surfaces with absorbing or diffusing material. Some absorption is normally needed to adjust the reverberation time of the space, but too much absorbing material in a control room can absorb too much signal energy. Diffusion does not have that limitation. Another approach to a reflection-free zone is that of orienting the reflecting surfaces so that reflections are directed away from the mixing position and toward the rear of the room.

Figure 15-3 shows an outline sketch of the front part of a control room. By splaying the side walls as shown, the early first-order reflections are directed toward the rear of the room instead of the mixing position as rectangular side walls would do. The early reflections are directed toward the rear of the room to be diffused and returned to the mixer as desired, providing helpful reverberatory sound. (The diffusers in the rear of the room are different from the diffusers that could be mounted at the front of the room on potentially reflecting surfaces.) The shaded area of Fig. 15-3 is a reflection-free zone of sufficient size to cover the console, the mixing engineer, and the producer's position behind the engineer.

A control room constructed to achieve this reflection-free zone and with proper diffusion at the rear wall is characterized by:

1. Clear sound, free from combing distortion.

2. An extremely wide "sweet spot."

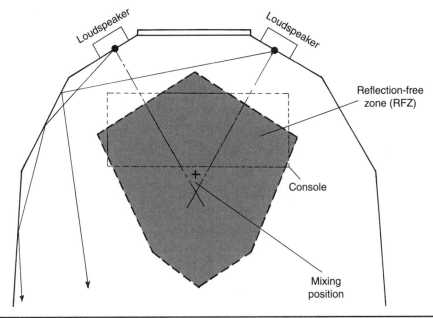

FIGURE 15-3 Creation of a reflection-free zone (RFZ) by splaying the side walls. The control room mixing position would be placed in the RFZ.

3. A spatial impression of being enveloped by the sound.

4. Accurate perception of the stereo image.

5. Flat low-frequency response with uniform modal decay.

6. Reproducibility and transferability of audio product.

Loudspeaker Mounting

Whether or not one is building an RFZ room, one source of early reflections should not be overlooked. The edges of a stand-alone loudspeaker cabinet can radiate sound through a wide angle because of diffraction from the cabinet edges. This diffracted sound can contribute to the early sound problem directly and by reflections from the wall behind the loudspeakers. By making the faces of the loudspeaker cabinets flush with the wall in which they are set, the diffraction effects and early reflections from this source can be eliminated. To achieve this, the loudspeakers can be mounted in soffits, so the faces of the monitors are flush with the angled wall.

Two-Shell Control Rooms

Control rooms are often built using a two-shell plan. Only a lightweight surface is necessary to deflect the mid- to high-frequency early sound in the proper direction (as in an RFZ design). Therefore, the inner shell of a control room built to a shape required for control of early sound can be of lightweight construction. Low-frequency energy, with its very long wavelengths, tends to permeate the space and must be dealt with independently. Thus the outer shell is heavy, and is built to control the low-frequency normal modes and to shield the inner room from environmental noises and likewise isolate the control room from the studio. The outer shell is often constructed of concrete blocks, normally with voids filled with sand or well-rodded concrete. The shape of the outer shell is somewhat controversial; some designers favor a trapezoidal shape, others prefer a rectangular shape. A trapezoidal shape does nothing toward reducing modal problems, it only shifts peaks and nulls to asymmetrical positions. A rectangular outer shell, at least, yields predictably symmetrical modal patterns.

Design Criteria

A professional-quality control room is a very expensive structure. If sufficient funds are not available, compromises must be made. Several less-than-perfect (but still good) control rooms are presented below and areas of compromise will be discussed. Control rooms are generally large, perhaps 20 to 25 ft across at the widest point, but have fairly short reverberation times. For example, a broadband reverberation time of about 0.3 sec is typical. Longer reverberation times could mask details in the monitored sound.

At first glance, a survey of control rooms may suggest that they employ very different designs. It is true that LEDE control rooms use fundamentally different acoustics than non-LEDE control rooms. But beyond this, many of the differences among control rooms are cosmetic. In fact, most control rooms share many design criteria. A control room must be quiet; for example, NCB of 15 to 20 is typical. This is relatively stringent, but compared to a studio with open microphones, a control room can tolerate some noise. On the other hand, a control room must have a high degree of isolation from other rooms. In particular, because of the open microphones in the studio, and potentially loud monitoring levels in the control room, the wall between them must provide a high degree of isolation. The large window in this wall must be designed and constructed with great care to ensure a high transmission loss. To a lesser extent, noise from adjoining rooms such as a machine room must be minimized as should noise intrusion from the control room to neighboring rooms such as offices.

Most control rooms are designed with axial symmetry so that the left and right halves of the room are identical. This is important to ensure proper mixing of the stereo or surround panorama. Control rooms often have bass traps or other means to control low-frequency modes, and thus provide a more consistent low-frequency response. Other broadband absorption is used to provide a moderate reverberation time. Diffusers are often mounted on the side and rear walls. Many, but not all, control rooms use a reflection-free zone so that first-order reflections do not pass through the mixing position. In most cases, an expansion ceiling (sloping upward from the monitors) is used, as opposed to a compression ceiling (sloping downward).

Diffusers are commonly used on the rear walls of control rooms; this is especially true when the front end of the room is absorptive. In most cases, rear-wall diffusers work well at providing an immersive sound field. However, some designers argue that a totally diffuse rear wall contributes too much ambience and thus can degrade imaging. Therefore, in some designs, some absorption is placed in the middle of the rear wall, with diffusers on either side. This alternative design, and ones like it, should be particularly considered when a surround-sound monitoring system is installed. Generally, surround-sound playback requires a more balanced front/back distribution of reflecting, absorbing, and diffusing elements in the control room.

Monitor loudspeakers are often flush mounted in soffits; in other cases, near-field monitors are placed on the console meter turret. In either case, loudspeakers are angled, so the center lines intersect at a point about 2 ft behind the mixing engineer's head; this is done so that the center lines pass through the ears. The front stereo loudspeakers are placed at angles of ±30°. Surround speakers can be placed in a variety of positions, and mounting should allow a variety of standard configurations; for example, placement at ±110° to ±120° is often used.

Design Example A: Control Room with Rectangular Walls

In this first control room example shown in Fig. 15-4, a rectangular control room is located in an existing structure. To avoid losing too much signal energy, absorbing material is minimized in favor of diffusing elements. The loudspeakers (1) are aimed in the normal stereo playback configuration. The Abffusor (manufactured by RPG Diffusor Systems, Inc.) is both a diffuser and an absorber (see Fig. 16-12). Abffusor panels (2) are mounted on the ceiling between the loudspeakers and the listening position. Additional Abffusor panels (3) are mounted on the left and right walls to intercept the reflections that would go toward the mixing position.

The mixing engineer is located between two parallel wall surfaces. Flutter echoes between the left and right walls would constitute a significant acoustical defect. These could be defeated with absorbent (which we wish to avoid). FlutterFree panels (4) (manufactured by RPG Diffusor Systems, Inc.) of surface-mounted hardwood molding are installed on both walls (see Fig. 4-17). These strips provide high-frequency diffusion that will eliminate a tendency toward flutter. The diffuser group at the

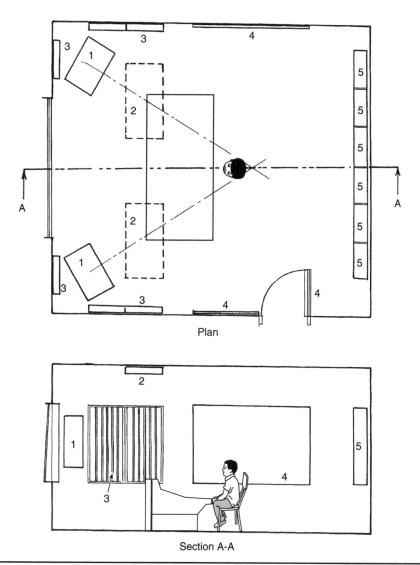

Plan

Section A-A

Figure 15-4   Control room example A: Rectangular side walls.

rear of the control room is composed of six 2 × 4-ft QRD-734 diffusers (5) (manufactured by RPG Diffusor Systems, Inc.), half of them with vertical wells and half with horizontal wells so as to give diffusion in both the horizontal and vertical planes (see Fig. 4-12).

Design Example B: Double-Shell Control Room with Splayed Walls

The second control room example, shown in Fig. 15-5, controls early reflections with a combination of room shaping and absorption. It has a shape that would fit well into a rectangular, heavy outer shell. The loudspeaker's (1) center lines intersect close behind

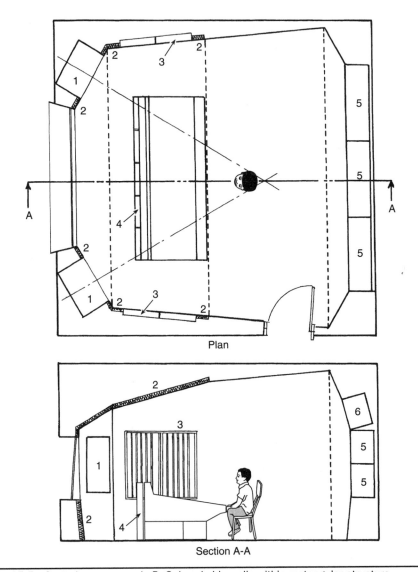

Plan

Section A-A

Figure 15-5 Control room example B: Splayed side walls within rectangular structure.

the mixing engineer's head. The side walls are splayed sufficiently to eliminate flutter echo. If this side-wall splaying is not sufficient to cope with the early-reflection path, Abffusor panels (3) can be set into the side walls.

The ceiling, side, and front walls are covered with absorbers (2), which are composed of a fabric-covered graduated-density glass fiber. This covering extends toward the rear as far as the armrest of the mixing console. The back of the console (4) and the wall under the window constitute a quasi-cavity, which can resonate at its own frequency. To control this resonance and to avoid "bass buildup," absorbers (4) are affixed to the rear of the console. They work with the absorbers mounted under the window.

The rear-wall diffuser array is made up of two rows of three QRD-734 diffusers (5) set horizontally with their wells vertical. On top of them is one row of three QRD-734 diffusers (6) set horizontally with their wells horizontal; these three diffusers are angled forward. This combination gives good diffusion in both the horizontal and vertical planes. As noted above, in some room designs, some of the rear-wall diffusion is reduced.

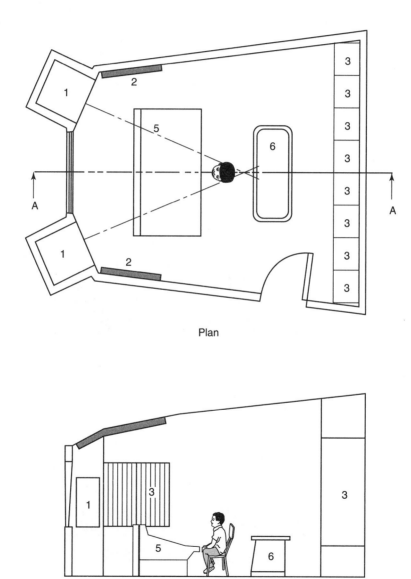

Plan

Section A-A

Figure 15-6 Control room example C: Splayed side walls.

Design Example C: Single-Shell Control Room with Splayed Walls

The third control room example is shown in Fig. 15-6. Elements of similarity to the previous two examples are evident, with a few significant differences. This room features slightly splayed side walls at a single angle. The splaying is not enough to eliminate the need for side-wall absorption/diffusion, but it is enough to eliminate flutter echo between the side walls. A producer's desk (6) has been located behind the mixing position. The monitor loudspeakers are recessed into soffits and mounted flush. The Abffusors (2) are mounted on the side walls to intercept the first-order reflections aimed at the mixing position. Other Abffusors are mounted on the underside of the ceiling to do the same thing. Alternatively, other types of diffusers could be used.

Because the rear of the room has space to accommodate a sizeable unit, a large diffuser array (3) is placed there. A variety of diffusers could be employed. A possible candidate is an array of Diffractal units (manufactured by RPG Diffusor Systems, Inc.). This design mounts diffusers within diffusers (using the fractal principle described in

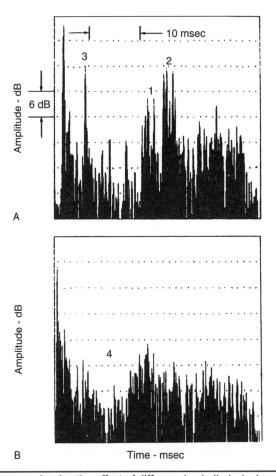

FIGURE 15-7   An echogram showing the effect of diffusers in similarly-designed control rooms. (A) Response without rear-wall diffusers. (B) Response with rear-wall diffusers. (*D'Antonio, 1984*)

Chap. 4) to provide very broadband diffusion. Another possibility is to build a low-frequency diffuser from ordinary concrete blocks, or to use DiffusorBlox (manufactured by RPG Diffusor Systems, Inc.). Diffusers reaching down to 100 Hz would certainly address potential modal problems in the room.

The effect of the rear-wall diffuser is clearly shown in Fig. 15-7. These records were taken in two very similar control rooms, the top one without rear diffusers and the bottom one with rear diffusers. In the top record, the early energy is confined primarily in the groups of specular reflections (1) and (2) arriving at about 17 and 21 msec, respectively. The reflection from the console (3) is also very prominent. The lower record shows the effect of the rear-wall diffuser. The valley of low reflections (4) is the initial time-delay gap of about 17 msec; after that come the highly diffuse reflections that add body and ambience to the direct signal. As noted, in some room designs, some of the rear-wall diffusion is reduced and replaced by absorption.

CHAPTER 16

Announce Booth

Historically, announce booths are among the earliest commercial acoustical spaces. They originated in the days of early radio broadcasts when they were used for on-air identification announcements. Announce booths are very small; they are literally booths. As we will see, their small size presents some particular acoustical challenges.

Today, announce booths are found in most radio and TV broadcasting studios. For example, a roving reporter and companion sound/camera operator cover an outdoor event, recording the reporter on a portable device. Returning to the studio, the producer calls for additional narration by the reporter to cover scenes in which the reporter does not appear. These post-production segments of the reporter's voice are recorded in an announce booth. If the booth is only a closet with some acoustical tile or carpet foam as acoustical treatment, the sound in live and post scenes may not match. In particular, the ambient information added by the booth's acoustics influences the sound of the post recording, degrading it and making it noticeably different from the live sound. When edited together, the result may be unacceptable.

As a demonstration to further illustrate the problem, record a voice outdoors, then with identical equipment record the same voice in a bathroom. It is easy to hear the difference, and it would be difficult or impossible to make the two sound the same. For example, equalization will rarely succeed at matching two very different recordings, unless they are both over-equalized to a consistent but poor sound.

Design Criteria: Isolation

As with any recording studio, announce booths must be acoustically isolated, at least in the speech-frequency range from 100 to 4000 Hz. The noise levels surrounding the booth will dictate what degree of insulating construction is needed. When the ambient noise level is low, and because very low-frequency insulation from external sources is not needed, simple gypsum-board stud walls may suffice. More sophisticated gypsum-board walls can also provide excellent insulation in the speech-frequency region. Noise outside the speech frequency region such as low-frequency HVAC vibration can be eliminated with electrical filters in the recording signal path. Because of the low volumes in an announce booth, only very modest insulation is needed to prevent noise intrusion to external spaces. For example, several announce booths could be built adjacent to one another and not suffer from interference.

The Small-Room Problem

The sound field in a small room (such as an announce booth) is dominated by resonances, which are completely absent outdoors. To minimize the variables in the comparisons to follow, a room having the dimensions of 6 × 8 ft with an 8-ft ceiling height will be used for each of the three examples in this chapter. Table 16-1 tabulates all the frequencies of the axial modes of this 8 × 6 × 8-ft space to 300 Hz beyond which modal colorations of the sound are minimal. To simplify the normal mode considerations of this small room, only the axial modes of the first order are calculated.

The lowest resonant frequency of this room is the axial mode associated with the longest dimensions, the 8-ft length and height of the room. The room's length forms the (1, 0, 0) mode and the height forms the (0, 0, 1) mode, both at 71 Hz; this is the lowest-frequency mode in the room. There will be no resonant support for sound of lower frequency than this. Because the length and the height of the room are the same at 8 ft there will be coincidences at 71 Hz and at every multiple of 71 Hz. The difference column shows coincidences at 71, 141, and 212 Hz, and a triple coincidence at 282 Hz. The effect of the modes at these frequencies will be stronger. Voice colorations may be encountered at these frequencies. The triple coincidence at 282 Hz is at a frequency that is high enough to lead us to hope for a minimal coloration problem. Such coincidences could have been avoided (or at least minimized) by choosing more favorable room proportions. Two identical dimensions, and dimensions that are multiples of each other, should be avoided.

Although the mean (average) difference between adjacent axial modal frequencies is 25.7 Hz, the gaps shown in the difference column in Table 16-1 suggest that the room will

	Length $L = 8$ ft $f_1 = 565/L$	Width $W = 6$ ft $f_1 = 565/W$	Height $H = 8$ ft $f_1 = 565/H$	Arranged in Ascending Order	Difference
f_1	70.6 Hz	94.2 Hz	70.6 Hz	70.6 Hz	0 Hz
f_2	141.3	188.3	141.3	70.6	23.6
f_3	211.9	282.5	211.9	94.2	47.1
f_4	282.5	376.7	282.5	141.3	0
f_5	353.1		353.1	141.3	47.0
				188.3	23.6
				211.9	0
				211.9	70.6
				282.5	0
				282.5	0
				282.5	70.6
				353.1	
				Mean difference: 25.7	
				Standard deviation: 27.4	

TABLE 16-1 Axial Modes in an Announce Booth (Room Dimensions: 8 × 6 × 8 ft)

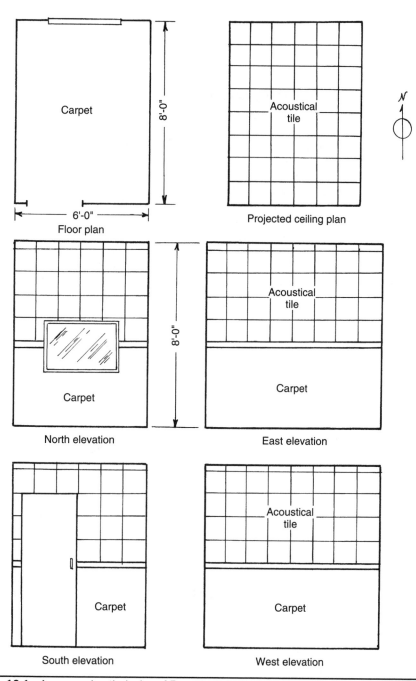

Figure 16-1 Announce booth design of Example A: Traditional treatment plan. This design illustrates the problems in a traditional booth; it is not recommended as a buildable design.

have a very irregular response below 300 Hz. This means that extra absorption must be provided to control these axial modes. Rooms smaller than 6 × 8 × 8 ft have even greater problems, as the axial modal frequencies are higher, but still able to degrade sound quality.

Design Example A: Traditional Announce Booth

This first example serves only to illustrate the acoustical problems in a traditional announce booth. It is not a studio to be built. Of course, existing traditional booths are very different, and presenting one to represent them all is a fragile fiction. However, there is reason to believe that acoustical tile and carpet are common to a great many existing booths, and these are featured in this example.

This assumed traditional talk booth is illustrated in Fig. 16-1. It is a small 8 × 6 × 8-ft rectangular space with common 2 × 4 stud framing covered on both sides with 1/2-in gypsum board. Despite all its deficiencies, because of its interior location inside a larger room, this structure should provide adequate insulation from outside noise except in extreme cases. The ceiling is covered with acoustical tile, as well as the walls above the wainscot strip. A heavy carpet with 40-oz pad covers the floor and the carpet (without the pad) is run up the wall to the wainscot strip.

Design Example A: Axial Modes

The sound pressure distribution of the axial modes in this 8 × 6 × 8-ft room is illustrated in Fig. 16-2. The small pressure graphs are moved outside the space to avoid confusion. The left portion of the figure shows the resonance between the east and west walls

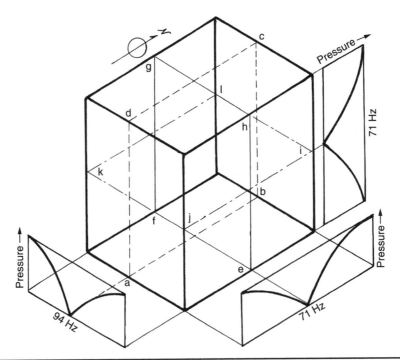

Figure 16-2   An axial-mode pressure diagram for the example 8 × 6 × 8-ft room.

creating the width mode (0, 1, 0). When this resonance is excited, the sound pressure over the entire east and west wall surfaces is high, with a null plane a, b, c, d, a extending from floor to ceiling.

The lower right portion of Fig. 16-2 shows the lengthwise mode (1, 0, 0). When this resonance is excited, the sound pressure is maximum over the entire surfaces of the north and south walls. There is a null plane e, f, g, h, e bisecting the room.

The upper right portion of Fig. 16-2 shows the vertical mode (0, 0, 1) at resonance between the floor and the ceiling. When this resonance is excited, sound pressure is high over the entire floor and ceiling surface with a null plane i, j, k, l, i bisecting the room.

These pressure maxima and minima dominate the acoustics of the space. The announcer could, for example, sit with his or her head positioned in or near all three null planes. The response of the microphone would fluctuate considerably as it is moved about in search of the best-sounding position. What the announcer hears and the microphone "hears" may be quite different. The announcer should use headphones.

Design Example A: Reverberation Time

A reverberation-time calculation is helpful in studying the frequency distribution of absorbence in a room. Table 16-2 shows the calculations applying to the booth's carpet and acoustical tile. The absorption $A = S\alpha$, where S is the surface area and α is the absorption coefficient. The result shows a long reverberation time below 250 Hz, as shown in Fig. 16-3 (solid line). However, other elements of the room will add additional absorption. In particular, the gypsum board and wood floor provide absorption in this needy region. It is easy to overlook this "free" absorption in a room, but its effect is very important.

The gypsum-board panels and the wood floor will contribute diaphragmatic absorption. The relatively large areas of wall panels and floor mean that their resonance points and peaks of absorption will be at low frequencies. Absorption at low frequencies is just what is needed to control the axial modes in this small room, and to minimize the great variations of sound pressure in the room caused by the axial modes (see Fig. 16-2). With the diaphragmatic absorption effective in the 70- to 250-Hz region, the modal maxima would be decreased and the null areas would be raised, but this would not completely overcome the uneven response caused by the axial modes.

By considering the absorption due to the walls and the floor in addition to that of the carpet and acoustical tile, the low-frequency reverberation time is lowered materially as shown in Fig. 16-3 (dashed line). However, without depending too much on the precision of the calculated reverberation time, the 0.1 sec of Fig. 16-3 is very low. A deadness may characterize this room; it may possibly have too much absorption for acoustical comfort and effectiveness. Another issue to consider is diffusion. As with any room, the acoustics are improved when the spatial distribution of sound energy is more evenly spread through the room. Diffusion treatment would be needed.

In summary, we can consider the various issues that would characterize the traditional announce booth of Example A:

1. The axial modes would produce an irregular frequency response in the sound field.

2. The diaphragmatic absorption of walls and floor may help to partly smooth the response.

Description	Area, ft²	125 Hz α	125 Hz Sα	250 Hz α	250 Hz Sα	500 Hz α	500 Hz Sα	1 kHz α	1 kHz Sα	2 kHz α	2 kHz Sα	4 kHz α	4 kHz Sα
Carpet, heavy 40-oz pad	140	0.08	11.2	0.24	33.6	0.57	79.8	0.69	96.6	0.71	99.4	0.73	102.2
Acoustical tile 1/2 in	155	0.07	10.9	0.21	32.6	0.66	102.3	0.75	116.3	0.62	96.1	0.49	76.0
Absorption, sabins			22.1		66.2		182.1		212.9		195.5		178.2
Reverberation time, sec			0.85		0.28		0.10		0.09		0.10		0.11
Drywall	224	0.29	65.0	0.10	22.4	0.05	11.2	0.04	9.0	0.07	15.7	0.09	20.2
Floor, wood	48	0.15	7.2	0.11	5.3	0.10	4.8	0.07	3.4	0.06	2.9	0.07	3.4
Total absorption, sabins (carpet + tile + drywall + floor)			94.3		93.9		198.1		225.3		214.1		201.8
Reverberation time, sec			0.20		0.20		0.10		0.08		0.09		0.09

TABLE 16-2 Reverberation-Time Calculations for a Traditional Announce Booth (Example A)

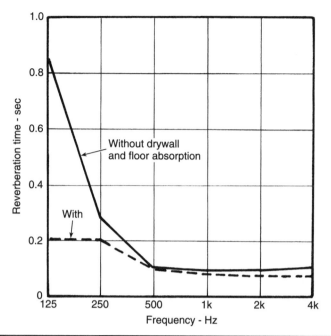

FIGURE 16-3 Graph of reverberation time for Example A.

3. Excessive mid- to high-frequency absorption would make the room too dead.

4. The booth lacks diffusion that is needed to thoroughly distribute the sound.

Design Example B: Announce Booth with TubeTraps

The second example of an announce booth is shown in Fig. 16-4. This room design features 16-in quarter-round TubeTraps (manufactured by Acoustic Sciences Inc.) in the four corners and 9-in half-round TubeTraps on four walls and the ceiling. These alone constitute the acoustical treatment of the room. The half-rounds on walls and ceiling provide absorption and, in conjunction with the strips of reflective wall surface between the TubeTraps, also provide diffusion. The perspective drawing of the announce booth with two walls removed, as shown in Fig. 16-5, might help in understanding the conventional drawings of Fig. 16-4.

The half-round TubeTrap is constructed as shown in Fig. 16-6. It is a rigid, easily handled unit with a fabric cover. The sound absorption characteristics of the various forms of the TubeTrap are shown in Fig. 16-7. Note that in this figure the absorption per linear foot is given directly in sabins, or absorption units, rather than the usual area and absorption coefficient approach.

In the typical studio, early reflections from bare areas of floor, walls, and ceiling are dominant and the source of problems. These reflections interact with each other, creating comb-filter coloration of the sound. In this design example, there are many discrete reflections from the strips of wall between the TubeTraps. In fact, there are so many of these reflections, offset from each other by small increments of time, that the cloud of comb filters produced are not audible as coloration but instead as a pleasant ambience. In other words, the density of reflections yields an overall smooth response.

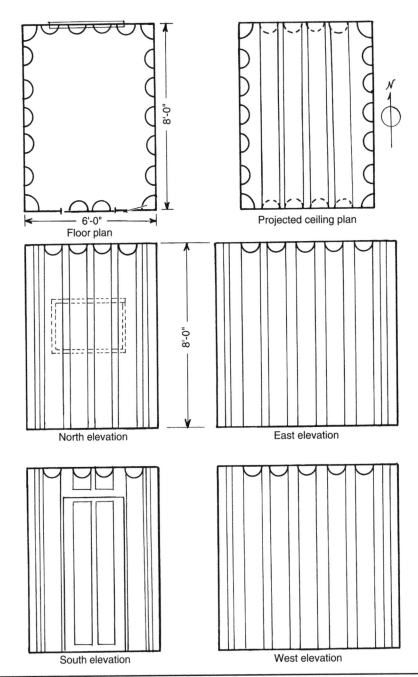

6'-0"
8'-0"
Floor plan

Projected ceiling plan

N

8'-0"
North elevation

East elevation

South elevation

West elevation

Figure 16-4 Announce booth design of Example B: Treatment plan using TubeTraps.

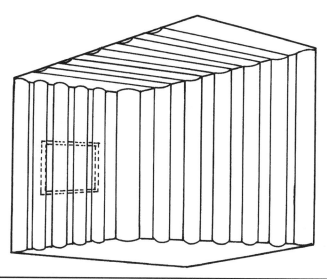

FIGURE 16-5 A perspective sketch for Example B.

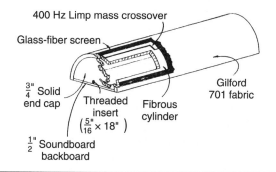

FIGURE 16-6 TubeTrap construction details.

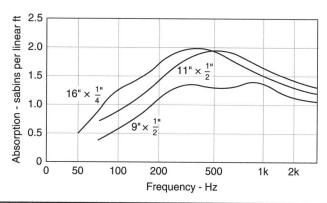

FIGURE 16-7 TubeTrap absorption.

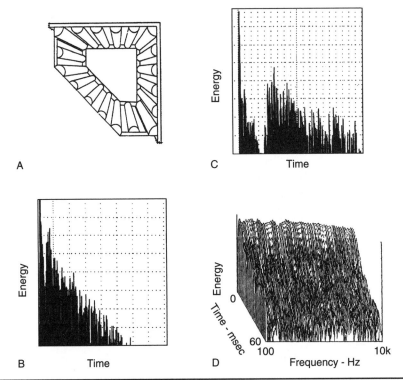

Figure 16-8 Analysis of an announce booth using TubeTraps. (A) Floor plan. (B) TEF measurement from 0 to 80 msec. (C) TEF measurement from 0 to 20 msec. (D) TEF "waterfall" plot.

Design Example B: Techron TEF Measurements

Figure 16-8(A) shows a small room that was constructed on the same principles as the announce booth in this example (see Fig. 16-4). Furthermore, time-energy-frequency (TEF) measurements were made in this room. The TEF measurements give a detailed view of the acoustical functioning of the room. In Fig. 16-8(B), the spike on the extreme left is the arrival of the direct sound at the microphone; the time scale runs from 0 to 80 msec. The uniform decay reveals a reverberation time of about 0.06 sec (60 msec). Figure 16-8(C) shows an expanded view of Fig. 16-8(B), with time running from 0 to 20 msec. There is a 3-msec initial time-delay gap between the arrival of the direct sound and the arrival of the stream of reflections from the TubeTrap grids on the walls and ceiling. The "waterfall" plot of Fig. 16-8(D) reveals a very smooth, consistent, and quite diffused decay throughout the audible band. These plots demonstrate that the acoustical treatment of the room yields a response that is smooth over both time and frequency.

Design Example B: Reverberation Time

The reverberation time calculations for the announce booth in this example are shown in Table 16-3, and the results are plotted in Fig. 16-9. The absorption $A = S\alpha$, where S is the surface area and α is the absorption coefficient. The very low calculated reverberation

Description	Length, ft	125 Hz A/ft	125 Hz A	250 Hz A/ft	250 Hz A	500 Hz A/ft	500 Hz A	1 kHz A/ft	1 kHz A	2 kHz A/ft	2 kHz A	4 kHz A/ft	4 kHz A
9-in TubeTraps, half-round	156	0.8	124.8	1.3	202.8	1.3	202.8	1.4	218.4	1.1	171.6	1.0	156.0
16-in TubeTraps, quarter-round	32	1.4	44.8	1.9	60.8	1.9	60.8	1.6	51.2	1.3	41.6	1.1	35.2
	Area, ft²	α	Sα	α	Sα	α	Sα	α	Sα	α	Sα	α	Sα
Drywall	224	0.29	65.0	0.10	22.4	0.05	11.2	0.04	9.0	0.07	15.7	0.09	20.2
Floor	48	0.15	7.2	0.11	5.3	0.10	4.8	0.07	3.4	0.06	2.9	0.07	3.4
Total absorption, sabins			241.8		291.3		279.6		282.0		231.8		214.8
Reverberation time, sec			0.078		0.065		0.067		0.067		0.081		0.088

TABLE 16-3 Reverberation-Time Calculations for an Announce Booth with TubeTraps (Example B)

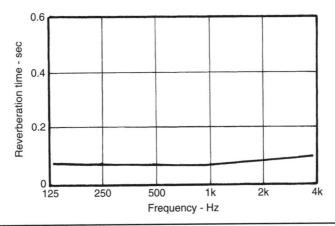

FIGURE 16-9 Graph of reverberation time for Example B.

time of about 70 msec approximately agrees with the measured TEF decay of Fig. 16-8(B) and is what is expected in a room such as this. A similarly low reverberation time was calculated for the traditional booth (see Fig. 16-3), which was overloaded with carpet and acoustical tile. The difference is that the dead characteristics of the TubeTrap room are improved with highly diffused sound.

What is the human perception within such a dead space? The announcer sitting in such a dead booth would have little acoustical feedback from the room. The announcer's own voice would sound somewhat unnatural to himself or herself. This can be rectified by providing the announcer with headphones and a quality playback of their own voice.

Design Example C: Announce Booth with Diffusers

The first announce booth example used absorption excessively. The second example used TubeTraps and would yield satisfactory recording quality of voice signals in a small room. This third example uses diffusing elements to also produce satisfactory small-room conditions.

The plan shown in Fig. 16-10 is based on the same 8 × 6 × 8-ft room dimension. The T-bar suspended ceiling is filled with twelve 2 × 2-ft Formedffusor quadratic-residue units (4) (manufactured by RPG Diffusor Systems, Inc.) as described in Chap. 4. These are not only diffusing elements, placed in the drop-ceiling, they also offer good mid- to low-frequency absorption as well.

To control axial modes, two Modex Corner bass traps (1) (manufactured by RPG Diffusor Systems, Inc.) are mounted in each corner (total quantity of 8) where modal pressures are highest. These units contain a membrane with high internal losses mounted in a trapezoidal cavity that fits into 90-degree corners; they measure 2 × 2 × 1 ft. The traps are available in tuning frequencies of 40, 63, and 80 Hz. This design uses 80-Hz units; the absorption measurements are shown in Fig. 16-11. These bass traps provide absorption at the dominant axial modes of the room at 71 and 94 Hz. As an 8-ft ceiling height is widely used in smaller studios, the 71-Hz normal mode associated with that dimension receives strong attenuation.

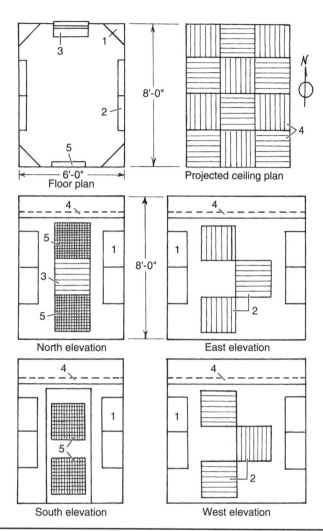

FIGURE 16-10 Announce booth design of Example C: Treatment plan using diffusers.

Abffusor units (2) (manufactured by RPG Diffusor Systems, Inc.) are mounted on the east and west walls, three to a side. These are positioned to minimize chance of flutter echo and to distribute the diffusion effect. These units measure 2 × 2 ft and have a broadband absorption characteristic, as shown in Fig. 16-12.

Two 2 × 2-ft Skyline units (5) (manufactured by RPG Diffusor Systems, Inc.) are mounted on the door in the south wall. Two others are mounted above and below the observation window in the north wall. These have maximum diffusion in two dimensions with a minimum of absorption. These diffusing surfaces are described in Chap. 4.

The only remaining diffuser is mounted over the observation window of the north wall. This 2 × 2-ft Diviewsor (3) (manufactured by RPG Diffusor Systems, Inc.) is a quadratic-residue diffuser based on the prime 7 sequence. It is a conventional type-734 diffuser, described in Chap. 4, except that the panels are of transparent Plexiglas. This allows visual communication through the glass window.

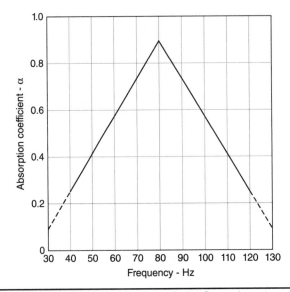

FIGURE 16-11   Absorption characteristic of an 80-Hz Modex Corner bass trap.

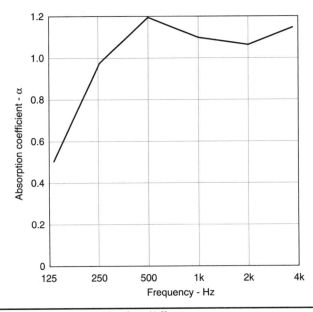

FIGURE 16-12   Absorption characteristic of an Abffusor.

Design Example C: Reverberation Time

Clearly, the announce booth in this example has a great deal of diffusion, but is there enough absorption to keep it from sounding too bright? A few calculations will answer this question. Table 16-4 shows the reverberation-time calculations for the announce booth with diffusing modules. The absorption $A = S\alpha$, where S is the surface area and α

Description	Area, ft²	125 Hz		250 Hz		500 Hz		1 kHz		2 kHz		4 kHz	
		α	Sα	α	Sα	α	Sα	α	Sα	α	Sα	α	Sα
Modex Corner (8)	32	0.18	5.8	0.1	3.2	0.07	2.2	0.05	1.6	0.03	1.0	0.02	0.6
Abffusors (6)	24	0.82	19.7	0.90	21.6	1.07	25.7	1.04	25.0	1.05	25.2	1.04	25.0
Formedffusors (12)	48	0.53	25.4	0.37	17.8	0.38	18.2	0.32	15.4	0.15	7.2	0.18	8.6
Drywall	224	0.29	65.0	0.10	22.4	0.05	11.2	0.04	9.0	0.07	15.7	0.09	20.2
Floor	48	0.15	7.2	0.11	5.3	0.10	4.8	0.07	3.4	0.06	2.9	0.07	3.4
Total absorption, sabins			123.1		70.3		62.1		54.4		52.0		57.8
Reverberation time, sec			0.15		0.27		0.30		0.35		0.36		0.33

TABLE 16-4 Reverberation-Time Calculations for an Announce Booth with Diffusers (Example C)

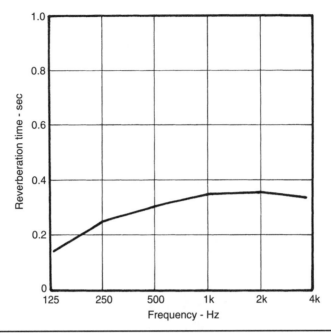

Figure 16-13   Graph of reverberation time for Example C.

is the absorption coefficient. No sound absorption material has been added to the room apart from that intrinsic to the diffusing elements and the structure itself. Both the Abffusor and Formedffusor units provide absorption in addition to their main function as diffusers. From the calculations of Table 16-4 come the reverberation-time data plotted in Fig. 16-13.

The reverberation time of approximately 0.3 sec is appropriate for such a small studio. It is suggested that anyone building a small studio like this and treating it with diffusing elements alone should listen analytically with the floor bare and then if desired, add a 5 × 7-ft rug or even carpet the entire floor to bring the reverberation time down to about 0.2 sec.

Example	Acoustic Treatment	Normal Mode Treatment	Reverberation Time at 500 Hz	Microphone Placement Sensitivity	Announcer Hearing Own Voice
A	Acoustical tile, carpet	*(only)	0.1 sec	Poor	Needs headphones
B	TubeTraps	*plus TubeTraps	0.07 sec	No problem	Needs headphones
C	Diffusers	*plus diffusers	0.3 sec	No problem	Natural

*Drywall and floor diaphragmatic absorption.

Table 16-5   Comparison of Treatment Examples A, B, and C

Design Example C: Evaluation

With so many diffusing units, the sound field in this example room should be thoroughly diffused. The announcer should experience a sense of being in a larger space, and the sound he or she hears should have a superior fullness and clarity. The microphone pickup should be relatively independent of microphone location and the sound of the voice should be quite natural.

Comparison of Design Examples A, B, and C

The same small studio ($8 \times 6 \times 8$ ft) has been treated in the traditional way and two contemporary ways using various acoustical treatment devices. The results are summarized in Table 16-5.

CHAPTER 17

Audio/Video/Film Workroom

I n the world of commercial acoustical design, much of the spotlight falls on recording studios and control rooms. However, much important work is carried out in more modest spaces. In particular, audio, video, film production, and postproduction work is done in small studios or workrooms. As with larger studios, these rooms demand accuracy in the sound field and comfortable working conditions. However, these needs must often be met on a very tight budget. As such, the design of these workrooms is a challenge to any designer.

Audio Fidelity and Near-Field Monitoring

In both large control rooms and audio/video/film workrooms, small loudspeakers placed on the instrument bridge of audio consoles supplement or replace large wall-mounted monitors. Such near-field monitors allow engineers to hear the sound "close up" and may give them a confidence they might not get from distant monitors. Some engineers prefer the smaller loudspeakers on the assumption that they help appreciate how the program will sound to consumers. Others use them a great proportion of the time out of necessity; with near-field monitors, the listener hears mainly direct sound; therefore, a room with poor acoustics will have less effect. Another advantage to small loudspeakers is their small size; in many workrooms, there simply isn't enough space to wall-mount large loudspeakers and be seated the required distance away from them.

The question of sound quality is not only a matter of one loudspeaker's specifications against another, it is also a matter of room interaction. In particular, early reflections can add distortion to what the engineer hears from large loudspeakers, but this can be avoided by listening to near-field loudspeakers. These early reflections are separated in time because they travel slightly different distances. The different reflections are replicas of the same signal with time shifts between them and will form comb filters when they combine; this results in distortion and coloration of the sound. The potentially superior sound of large monitor loudspeakers is distorted by the early reflections. Eliminating the reflections restores the superior sound at the listening position. Listening to near-field loudspeakers discriminates against the reflections, but accepts the potentially inferior sound quality of smaller loudspeakers.

In the design of this audio/video/film production workroom, priority is given to eliminating early reflections at the listening position so that wall-mounted monitors can be trusted to convey an accurate sound field. That is not as simple as it may seem; early reflections are easily generated, and require treatment to avoid their deteriorating effect. Of course, near-field monitors can additionally be used. In other words, both distant and near-field monitoring have advantages and disadvantages; the room design considered here allows for both types of monitoring.

Axial-Mode Considerations

The first step in minimizing the deleterious effect of axial modes is to select proper proportions for the room. This is discussed in Chap. 5, with details given in Table 5-4. Sepmeyer's C ratio, 1.00:1.60:2.33, is used in this workroom. Table 17-1 lists the axial-mode frequencies for this room. Listing these frequencies in ascending order and noting the difference in frequency between adjacent modes gives a good appraisal of possible modal problems. The mean (average) difference is 9.5 Hz, although the real differences vary from 0 to 20.2 Hz. A near-coincidence occurs at both 141 and 282 Hz, and a definite coincidence at 235 Hz.

Only the 141-Hz mode threatens a sound coloration; the others are high enough in frequency to avoid coloration effects. The 141-Hz coincidence is associated with both the 12-ft ceiling height and the length of the room. The chances of the 141-Hz coloration being audible are slight because the tangential and oblique modes (even though lower in energy) will tend to fill in the gaps between the axial modes. The advantage of a larger room like this is that problems of mode spacing tend to recede.

Monitor Loudspeakers and Early Sound

The general plan of the workroom is shown in Fig. 17-1. The volume of the room is 7068 ft^3, excluding the space behind the loudspeakers. A workstation with monitor loudspeakers and a small mixing console is placed along the front (north) wall. Diffusers and a workbench are placed along the rear (south) wall. To give the mixing engineer the best sound possible from the monitors, two soffits are centered on 60° lines converging slightly behind the listening position. This ensures that the direct sound from the monitors passes by the listener's ears. The function of the soffits is to place the faces of the monitors flush with the angled wall. The corners of a stand-alone loudspeaker cabinet radiate sound through a wide angle by diffraction from the corners. This diffracted sound from the loudspeaker edges contributes to the early sound problem directly and by reflections from the wall behind the loudspeakers. By making the faces of the monitor cabinets flush with the wall in which they are set, the diffraction effects and early sound from this source are eliminated.

Loudspeaker-cabinet diffraction issues can be minimized by flush-mounting the cabinets. However, the room creates other time-delay problems that must be addressed. A consideration of the favorable and unfavorable sound rays from the monitors is shown in Fig. 17-2. The solid lines indicate direct rays arriving at the listening position as desirable sound. The broken lines are reflected from the walls, floor, and ceiling. Because they arrive at different times, they result in comb-filter distortion. These reflections can be controlled by locating areas of absorbent material at the indicated positions on the side walls and on the ceiling over the head of the mixing engineer (see Fig. 17-8). Alternatively, the reflections could be controlled by placing diffusers in these same locations. For this design example, absorption patches will be used. Reflections from the floor are absorbed by a carpet. One early-arrival reflection comes from the face of the mixing console; this is unavoidable.

	Length $L = 27.96$ ft $f_1 = 565/L$	Width $W = 19.2$ ft $f_1 = 565/W$	Height $H = 12$ ft $f_1 = 565/H$	Arranged in Ascending Order	Difference
f_1	20.2 Hz	29.4 Hz	47.1 Hz	20.2 Hz	9.2 Hz
f_2	40.4	58.9	94.2	29.4	11.0
f_3	60.6	88.3	141.3	40.4	6.7
f_4	80.8	117.7	188.3	47.1	11.8
f_5	101.0	147.1	235.4	58.9	1.7
f_6	121.2	176.6	282.5	60.6	20.2
f_7	141.5	206.0	329.6	80.8	7.5
f_8	161.7	235.4		88.3	5.9
f_9	181.9	264.8		94.2	6.8
f_{10}	202.1	294.3		101.0	16.7
f_{11}	222.3			117.7	3.5
f_{12}	242.5			121.2	20.1
f_{13}	262.7			141.3	0.2
f_{14}	282.9			141.5	5.6
f_{15}	303.1			147.1	14.6
				161.7	14.9
				176.6	5.3
				181.9	6.4
				188.3	13.8
				202.1	3.9
				206.0	19.3
				222.3	13.1
				235.4	0.0
				235.4	7.1
				242.5	20.2
				262.7	2.1
				264.8	17.7
				282.5	0.4
				282.9	11.4
				294.3	8.8
				303.1	
				Mean difference: 9.5 Standard deviation: 6.4	

TABLE 17-1   Axial Modes in an Audio/Video/Film Workroom (Room Dimensions: 27.96 × 19.2 × 12 ft)

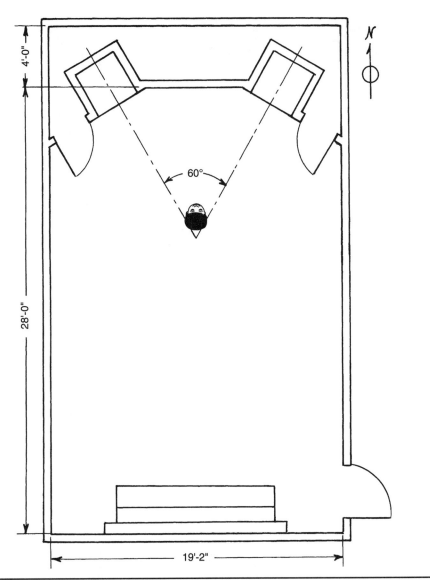

FIGURE 17-1 The audio/video/film workroom plan view.

Late Sound

Once the initial time-delay gap problems are minimized by the use of soffits and patches of absorbent or diffusers, the direct sound from the monitors should be free from coloration-producing early reflections. This direct sound allows accurate perception of the monitor playback, but it is only the first part of the complete sound field. The late sound, made up of reflections from the rear of the room, completes the sound field.

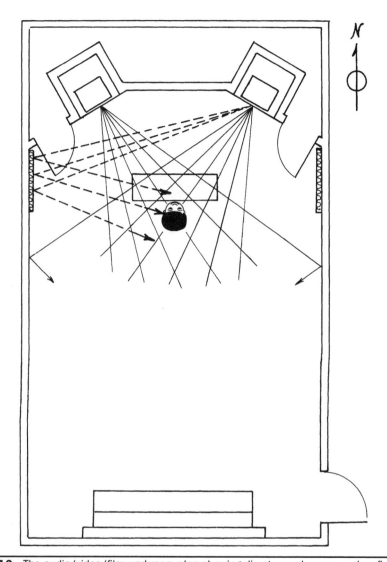

Figure 17-2 The audio/video/film workroom plan showing direct sound versus early reflections.

A view of both the early and the late sound is shown in Fig. 17-3. The direct sound reaching the ears of the engineer at the workstation establishes both the zero level and the zero time of this diagram. Following the arrival of the direct sound, a period of relative silence occurs as the sound passes the engineer on its way to the rear wall. This is not a completely silent period; there are always minor reflections of low level but they can be neglected. These miscellaneous reflections are down perhaps 30 dB, and hence not significant in the overall perception.

Using the room dimensions (see Fig. 17-1) we can estimate the level and time values of the late reflections. The direct sound must travel about 9 ft from the loudspeakers to the engineer, and depending on the path, must travel an additional 34 to 40 ft to return

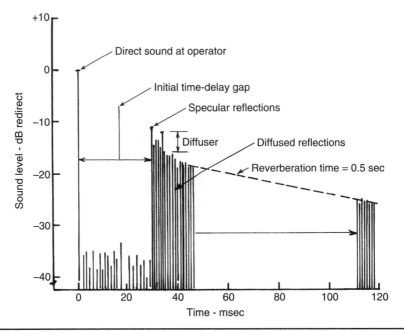

FIGURE 17-3 Early reflections and reverberation in the workroom.

to the engineer as reflections from the face of the rear workbench and the rear wall. The level of the reflection from the workbench will be about 20 log 9/34 = 11.5 dB, and the level of the reflection from the rear wall will be about 20 log 9/40 = 12.9 dB. These rough estimates, based on inverse-square law propagation, are accurate enough for present purposes.

The delays of these two rear reflections are about 34 ft/1130 ft/sec = 30 msec, and 40 ft/1130 ft/sec = 35 msec. Therefore, the reflection from the workbench is 11.5 dB at 30 msec of delay, and the reflection from the wall is 12.9 dB at 35 msec delay. These specular reflections are plotted in Fig. 17-3.

The rear-wall sound also falls on a bank of Skyline diffusers. The level of the return from the diffusers is 6 or 8 dB less than what falls on the diffuser face due to normal diffuser action. The flood of rear-wall reflections, both specular and diffused, decays at a rate determined by the reverberation time of the room. Our goal is a reverberation time of about 0.5 sec, and that is the slope drawn in Fig. 17-3.

A substantial portion of the sound falling on the rear wall hits the reflection phase-grating diffusers, and is diffused both in the horizontal and the vertical directions. The reflections from the workbench and the rear wall not covered with diffusers mix with the diffused sound and return to the engineer's ears delayed in time. The direct sound sweeps past the engineer to the rear workbench and wall and returns to the engineer's ears, arriving about 30 to 35 msec later than the direct sound. The initial time-delay gap in the room characteristic is very important to the engineer, enabling him or her to clearly hear the direct sound and also natural room ambience that does not degrade the accuracy of the direct sound. Moreover, because of the Haas effect, the engineer does not hear the ambience as a separate event, and he or she localizes the ambience as coming from the front of the room with the direct sound.

Reverberation Time

The volume of the room is 7068 ft^3; this volume exceeds the usual "small room" category and allows the room's acoustics to approach the condition of thoroughly mixed sound. Therefore, the concept of reverberation time can be applied accurately in this room. Table 17-2 shows the calculations of reverberation time of the room at standard frequencies. Starting with no specific room treatment, the major absorber is the 1140 ft^2 of gypsum board with its diaphragmatic absorption. As we see the gypsum-board absorption peak in the low-frequency region, it is natural to think of carpet (with its primary absorption in the higher frequencies) to compensate. This justifies placing carpet over the entire floor surface. The only other absorbing areas, and these are of minor effect because of their small size, are the two wall and one ceiling 4 × 4-ft panels to intercept the early reflections. The absorption of the gypsum-board surfaces, the carpet, and the small panels is computed in Table 17-2. The absorption $A = S\alpha$, where S is the surface area, and α is the absorption coefficient. The resulting reverberation times are plotted in Fig. 17-4.

The gypsum board brings the 250-Hz reverberation time down from a very high value of 1.18 sec, but it does not bring it down far enough. Compared to the desired reverberation time of 0.5 sec, the values above 500 Hz are not very far off. About 400 ft^2 of Helmholtz resonators tuned to 250 Hz would decrease this 1.18-sec peak to 0.5 sec, but this modification would be unwise at this stage. There are too many uncertainties in gypsum board and carpet absorption to be that specific. A wiser approach would be to wait until the structure is built and the carpet laid, and then to measure the reverberation time to see precisely what correction is needed. The correction will be relatively small.

A goal of a 0.5-sec reverberation time has been indicated, but the client might very well prefer 0.4 sec (a slightly dryer acoustic) or 0.6 sec (a slightly livelier acoustic). Such a preference and the measured values would provide the basis for the right correction.

Workbench

Figure 17-5 shows the south elevation of the room; the rear workbench is placed there, and the wall is largely covered by an array of 2 × 2-ft Skyline diffusing modules (manufactured by RPG Diffusor Systems, Inc.). This wall-mounted array is made up of 18 of the units shown in Fig. 17-6. These units diffuse sound in both horizontal and vertical directions.

Beneath this diffusing area a workbench is suggested. Any reflections from this bench will simply add to the rear wall specular and diffuse reflection mixture. If placed anywhere else in the room, reflections from the bench would clutter the initial time-delay gap with spurious reflections. Moreover, if placed along one wall, reflections from the bench would compromise the bilateral symmetry of the room's acoustics.

The engineer at the workstation may need an assistant to do the many jobs to keep the workstation efficiently occupied. Certain equipment will be necessary to make this assistant efficient. A shelf at the bottom of the diffusing panel is intended to hold such equipment. A few such pieces will have a negligible effect on the functioning of the wall diffusers.

Description	Area, ft²	125 Hz		250 Hz		500 Hz		1 kHz		2 kHz		4 kHz	
		α	Sα	α	Sα	α	Sα	α	Sα	α	Sα	α	Sα
Drywall	1148	0.29	332.9	0.10	114.8	0.05	57.4	0.04	45.9	0.07	80.4	0.09	103.3
Carpet, heavy 40-oz pad	589	0.08	47.1	0.24	141.4	0.57	335.7	0.69	406.4	0.71	418.2	0.73	430.0
Wall and ceiling panels, 703, 2-in flat on wall	48	0.24	11.5	0.77	37.0	1.13	54.2	1.09	52.3	1.04	49.9	1.05	50.4
Total absorption, sabins			391.5		293.2		497.3		504.6		548.5		583.7
Reverberation time, sec			0.88		1.18		0.77		0.69		0.63		0.59

TABLE 17-2 Reverberation-Time Calculations for an Audio/Video/Film Workroom

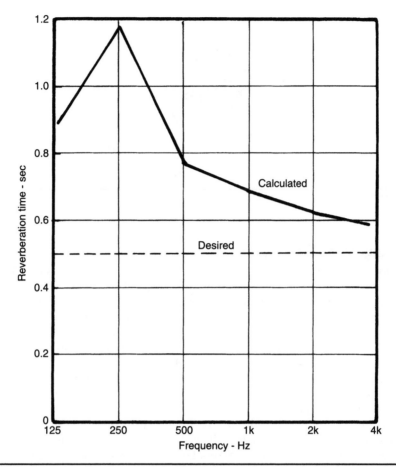

FIGURE 17-4 A reverberation time graph for the workroom showing calculated reverberation time, and the design goal.

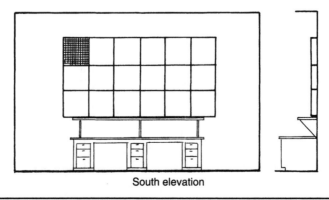

South elevation

FIGURE 17-5 South elevation of the workroom.

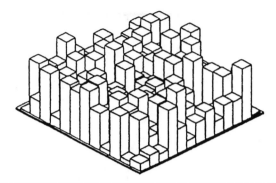

FIGURE **17-6** An isometric drawing of a Skyline diffuser.

Mixing Engineer's Workstation

Figure 17-7 shows the north elevation of the workroom. This drawing details the position of the mixing engineer to the monitor loudspeakers and the video display. The line of sight is high enough that there is room for equipment without blocking the view, but care must be exercised in this regard. In particular, the monitors and display should not be placed too high relative to the sitting position of the engineer.

Racks of auxiliary equipment can be mounted under the desk on either side of the engineer's feet. Unless absolutely necessary, another workstation should not be placed behind the engineer; equipment on it would send comb-filtering reflections back to the engineer's ears. Doors have been suggested to make the space behind the monitors available for storage.

Figure 17-8 presents a side view of the engineer's position along the west elevation. The relative position of the side wall and ceiling early-sound panels is shown.

Should the monitor loudspeakers be vertical, or inclined downward to place the engineer on the axis of the monitor? It would be best to be on-axis, but in high-quality

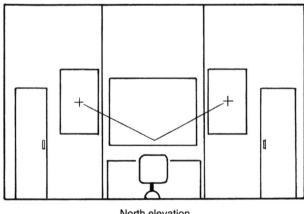

North elevation

FIGURE **17-7** North elevation of the workroom.

Figure 17-8 West elevation of the workroom.

monitors sound 10° off-axis is practically the same. The simplest procedure is to accept a vertical partition and 10° off-axis sound. The partition could be made in two sections, one inclined at the same horizontal and the same vertical angle as the face of the left loudspeaker, and the other inclined to coincide similarly with the face of the right loudspeaker. This would cause some minor problems with the doors and the video display, which could be worked out. By flush-mounting the loudspeakers, this source of comb-filter reflections can be avoided.

Lighting

It is suggested that the ceilings and upper walls be painted flat black, and that shaded light fixtures be hung from the ceiling. Track lighting to highlight the rear workbench and diffusers could provide sufficient working illumination, as well as a dramatic touch. Similar track lighting could provide working illumination at the workstation.

Background Noise Level

This room is intended primarily for postproduction work; hence recording of production sound would be rare. The primary need is that background noise be kept at an appropriately low level so that the engineer can evaluate sound from the monitors with no interference. Isolation is discussed in Chap. 6.

Video Display

A possible video display position is indicated in the north elevation of Fig. 17-7. A flat-panel display could be easily mounted on the front wall. Alternatively a video projector could be placed near the engineer's desk. The projector, video monitors, and other equipment on the desk are all possible culprits in producing early sound reflections that could distort the sound the engineer hears. As a last resort, absorbing blankets over the top and back of desk equipment might be necessary to control these reflections. Also, ambient noise from a projector's cooling fan must be minimized.

CHAPTER 18

Teleconference Room

Any corporate environment needs one or more conference rooms for employees and clients to meet. However, the expense of travel and loss of time required for face-to-face meetings are so great as to encourage other methods. Many types of audio and video communications systems are readily available to bring people together irrespective of distance and at far less cost than traveling.

The availability of audio and video systems is one part of the solution; also needed is an acoustically suitable space in which to use them. A dedicated space for in-house meetings and teleconferences can be a useful corporate resource, and can mean the difference between an unprofessional makeshift hookup and a professional teleconference meeting. This kind of teleconference room can thus connect people to another or many locations on a global basis while ensuring comfortable and natural speech intelligibility.

Design Criteria

Speech intelligibility is the single most important requirement in a meeting room of any kind. When speech is conveyed over data lines, intelligibility can be an even greater problem because of distortion at the remote location or in the transmission channel. Given a clear channel, and satisfactory local equipment, good room acoustics at both ends of the communication line will optimize speech intelligibility.

Speech is most intelligible in acoustically dead spaces. Conversely, intelligibility in highly reverberant spaces is very poor. However, adequate acoustical design of a teleconference room involves many things other than sound absorption. The background noise level must also be low. Specifically, it is well to keep the background noise level below the NCB-20 contour (see Fig. 11-7). To achieve this low level of background noise, attention must, among other things, be given to HVAC noise. A low-velocity HVAC system is necessary, along with wrapped and lined ducts, and avoidance of certain fittings such as noisy air diffusers. Nearby noisy operations external to the teleconference room and the sound attenuation of the walls of the space must be brought into conformance. Once these are cared for, attention is directed to the treatment of the inside of the teleconference room.

Shape and Size of the Room

The size of the teleconference room must be determined by the number of participants to be accommodated at one time. For 12 people plus the director, a 21-ft × 14-ft, 5-in × 9-ft space is selected for this design example; this fits the room proportions of 1.0:1.4:1.9.

	Length L = 21 ft $f_1 = 565/L$	Width W = 14.42 ft $f_1 = 565/W$	Height H = 9 ft $f_1 = 565/H$	Arranged in Ascending Order	Difference
f_1	26.9 Hz	39.2 Hz	62.8 Hz	26.9 Hz	12.3 Hz
f_2	53.8	78.4	125.6	39.2	14.6
f_3	80.7	117.5	188.3	53.8	9.0
f_4	107.6	156.7	256.1	62.8	15.6
f_5	134.5	195.9	313.9	78.4	2.3
f_6	161.4	235.1		80.7	26.9
f_7	188.3	274.3		107.6	9.9
f_8	215.2	313.5		117.5	8.1
f_9	242.1			125.6	8.9
f_{10}	269.0			134.5	22.2
f_{11}	296.0			156.7	4.7
				161.4	26.9
				188.3	0.0
				188.3	7.6
				195.9	19.3
				215.2	19.9
				235.1	7.0
				242.1	9.0
				251.1	17.9
				269.0	5.3
				274.3	21.7
				296.0	
				Mean difference: 12.9 Standard deviation: 7.83	

TABLE 18-1 Axial Modes in a Teleconference Room (Room Dimensions: 21 ft × 14 ft, 5 in × 9 ft)

A listing of the normal modes of this space is given in Table 18-1. The difference column appears reasonable, with a single degeneracy at 188 Hz. The chance of this degeneracy causing an audible coloration is slim because experience has shown that few colorations have been detected at this relatively high frequency. Furthermore the presence of tangential and oblique modes will tend to minimize the effect of the degeneracy. The volume of the room is 2722 ft³.

Floor Plan

Figure 18-1(A) shows the floor plan of the suggested teleconference room and many of the acoustical elements: loudspeakers (1); video display (2); Skyline diffusers (3); Modex Corner bass traps (4); low-frequency Helmholtz absorber (5); carpet (6).

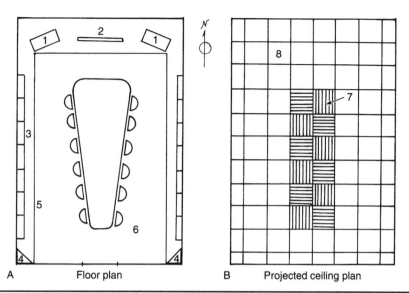

A   Floor plan   B   Projected ceiling plan

Figure 18-1   Floor and ceiling plans for a teleconference room.

A conference table with a wedge shape is suggested so that each seated participant has a reasonably good view of the video display, or any activity at the head of the table. The 30-in-high shelf at the front of the room supports loudspeakers as well as a flat-panel display. Alternatively, the loudspeakers could be wall-mounted, and a retractable projection screen could be installed in the ceiling. Sliding doors make the space underneath this shelf available for storage. The 30-in-high shelf continues along the sides of the room. To eliminate any possibility of flutter echoes between the east and west walls, and to provide supplemental diffusion for the room, 14 Skyline modules (manufactured by RPG Diffusor Systems, Inc.) are mounted on each side above the shelf. These are two-dimensional diffusers.

Ceiling Plan

The room uses a standard T-bar suspended ceiling with a 16-in airspace, as shown in Fig. 18-1(B). In the center of this frame, directly over the conference table, are 14 Abffusor units (7) (manufactured by RPG Diffusor Systems, Inc.), which both diffuse and absorb sound. The rest of the ceiling rack is filled with Tectum Designer Plus ceiling panels (8) (manufactured by Tectum Inc.) having the dimensions $24 \times 24 \times 1\text{-}1/2$ in. Each Tectum panel is scored into nine squares, which give a bold and interesting appearance. An overhead video projector can be mounted here if desired.

Elevation Views

The north, east, south, and west elevation sketches are shown in Fig. 18-2. These sketches help to relate the two types of diffusers. Because reverberation time must be very short, all available wall space between the 30 in-high shelf top and the suspended

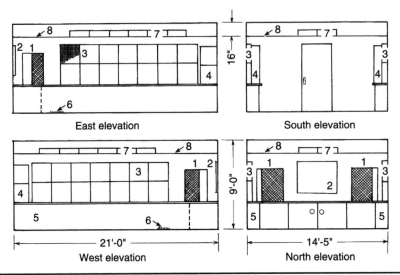

FIGURE 18-2 Four elevation views of a teleconference room.

ceiling is covered with Tectum wall panel on a D-20 mounting (mounted on 3/4-in furring strips). The Skyline modules (3) are cemented to this Tectum wall panel.

The south elevation is shown in Fig. 18-2. Two Modex Corner bass traps (4) (manufactured by RPG Diffusor Systems, Inc.) are placed on the 30-in-high shelf on either side of the door. These help control low-frequency effects caused by low-frequency modes. If such effects persist, two similar stacks of the same bass units should be placed in the corners behind the loudspeakers. It is not expected that additional units will be needed. Because speech intelligibility is the principal concern, the room's low-frequency response is not critical.

A partial sectional view through a wall is shown in Fig. 18-3. The 30-in-high shelf along the side walls is only about 12-in wide to avoid intruding on the limited space of the room. On both sides of the room, beneath this shelf, a Helmholtz resonator low-frequency absorber is mounted. Its purpose is described below.

Reverberation Time

Reverberation-time calculations for the teleconference room are shown in Table 18-2. The absorption units (sabins) are shown for six standard frequencies for the following elements: (a) 12 ceiling Abffusors, (b) ceiling Tectum Designer Plus panels, (c) Tectum wall panels, and (d) heavy carpet with pad. These calculations lead directly to the plotted graph of Fig. 18-4 labeled "without compensation."

The calculated reverberation time varies from 0.16 to 0.38 sec, which is a reasonable range. A value of 0.2 sec would represent a room with even greater absorption, which is just what is needed for good speech intelligibility. The room has adequate absorption at 125 Hz due to the diaphragmatic action of wall and ceiling gypsum board. The 0.38-sec peak at 250 Hz suggests the possibility of adding a peak of absorption at this frequency.

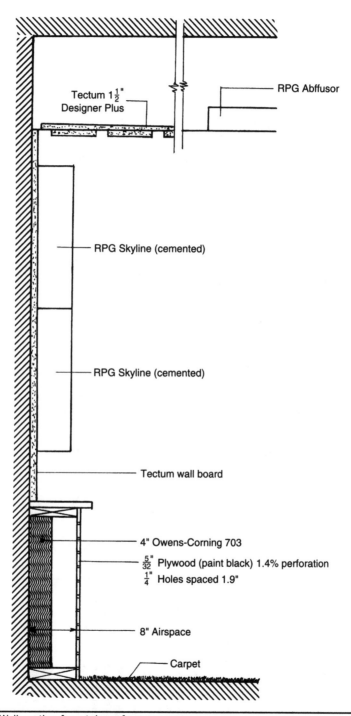

Tectum $1\frac{1}{2}$"
Designer Plus

RPG Abffusor

RPG Skyline (cemented)

RPG Skyline (cemented)

Tectum wall board

4" Owens-Corning 703

$\frac{5}{32}$" Plywood (paint black) 1.4% perforation
$\frac{1}{4}$" Holes spaced 1.9"

8" Airspace

Carpet

FIGURE 18-3 Wall section for a teleconference room.

Description	Area ft²	125 Hz		250 Hz		500 Hz		1 kHz		2 kHz		4 kHz	
		α	Sα	α	Sα	α	Sα	α	Sα	α	Sα	α	Sα
Drywall, 1/2 in on 16-in centers	940	0.29	272.6	0.10	94.0	0.05	47.0	0.04	37.6	0.07	65.8	0.09	84.6
Abffusor, ceiling	48	0.82	39.4	0.90	43.2	1.07	51.4	1.04	49.9	1.05	50.4	1.04	49.9
Tectum, ceiling	246	0.35	86.1	0.42	103.3	0.39	95.9	0.51	125.5	0.72	177.1	1.05	258.3
Designer Plus 24 × 24 × 1-1/2 in													
Carpet, heavy + pad	203	0.08	16.2	0.27	54.8	0.39	79.2	0.34	69.0	0.48	97.4	0.63	127.9
Tectum walls, D-20 mounting	360	0.07	25.2	0.15	54.0	0.36	129.6	0.65	234.0	0.71	255.6	0.81	291.6
Total absorption, sabins			439.5		349.3		403.1		516.0		646.3		812.3
Reverberation time, sec, w/o compensation			0.30		0.38		0.33		0.26		0.21		0.16
Helmholtz low-frequency compensation	97	0.80	77.6	0.90	87.3	0.68	66.0	0.28	27.2	0.18	17.5	0.12	11.6
Total absorption, sabins			517.1		436.6		469.1		543.2		663.8		823.9
Corrected reverberation time, sec			0.26		0.31		0.28		0.25		0.20		0.16

TABLE 18-2 Reverberation-Time Calculations for a Teleconference Room

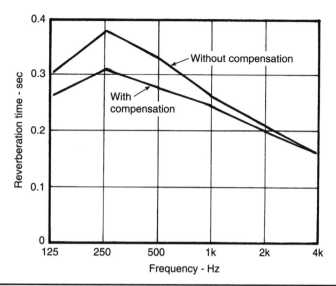

Figure 18-4 Reverberation time for a teleconference room.

For this reason, a Helmholtz unit having a peak about 250 Hz should be placed under the shelf. The 97 ft² available brings the reverberation time at 250 Hz down a bit, but not as much as desired. Even though it is impractical to equalize reverberation time to a reasonably flat 0.2 sec, reverberation time between 0.2 and 0.3 sec should make the room dead enough for good speech intelligibility. The two bass traps that are used to control very low-frequency room modes are not included in the reverberation-time calculation.

CHAPTER 19

Home Studio

Home studios are small recording and mixing spaces used by one or a few individuals. Given the sophistication and wide availability of recording hardware and software, a project studio can yield recordings of relatively high quality. The logical location for such a project studio is usually in the bedroom, basement, or garage. Musicians should not be discouraged by the prospect of recording in modest home studios. For example, the Foo Fighter's *Wasting Light* album was recorded in lead singer Dave Grohl's garage, and it received four Grammy awards.

With suitable acoustical treatment, and several good microphones and monitor speakers, a project studio can be used for both recording and mixing. Other equipment might comprise a small mixing console and peripheral gear, or simply a microphone preamplifier and a laptop. One drawback to most project studios is lack of isolation from surrounding rooms or the exterior. This means that sounds external to the studio can interfere with recording, and likewise performance and mixing can be intrusive to others.

The studio design presented here is somewhat unique. Floor space is at a premium in a home; many times there simply is not enough space to allow for a separate control room and recording room. Dividing a room to allow both purposes would result in separate spaces, but the spaces would be so small as to be useless for either function. In this room design, one room will be used for both recording and mixing. This dual-purpose studio is not ideal, but will be fully satisfactory for most purposes. The design is presented in three iterations, evolving from very low cost and progressing toward higher cost. The playback conditions are considered first, and then the suitability of the room for recording is discussed. Also, a design for a garage studio with separate control room and recording studio is presented.

Home Acoustics: Modes

All small rooms, including most bedrooms and garages, have modal resonances at audible frequencies. In particular, at low audio frequencies, the wavelength of sound is of the same order as the dimensions of the room. This means that sound-pressure peaks and nulls will arise as those frequencies are sounded in the music. This can degrade the low-frequency response of the room including the response of its reverberation. The effect of room modes can be decreased somewhat by choosing certain room proportions. But most home studios are built in existing rooms. With a room of predetermined dimensions, the only approach is sound treatment such as bass traps that will address resonant modes. While not a complete solution, treatment can provide very satisfactory acoustical results. The theory behind room modes is discussed in Chap. 5.

Home Acoustics: Reverberation

The average reverberation time of 50 living rooms has been reported by Jackson and Leventhall (Jackson and Leventhall, 1972) as being about 0.7 sec at 100 Hz, and decreasing to 0.4 sec at higher audio frequencies. This is quite a reasonable target range for a home studio. The average bedroom is smaller in size than the average living room and would have a somewhat lower reverberation time, but is still within a usable range. So, the reverberation times found in most homes are probably acceptable, or at least within a range that can be adjusted to be acceptable, for a home studio.

Home Acoustics: Noise Control

As for background noise, a typical home and neighborhood has a far higher noise level than one would like to have on a recording. Attempts to isolate the studio will almost certainly be insufficient. The only practical way around this is to record late at night when the world is quiet. The cost of isolation, and the interior volume it occupies, makes good isolation extremely difficult to achieve. This background noise problem has no easy solution, although working with close microphone placement will help.

It is also important to remember that recording and mixing, especially late at night, may be intolerable to others near the studio especially if amplified instruments are played. Many local ordinances limit residential noise levels particularly at night. For example, the noise-level limit at a property line might be 45 dBA, or 5 dB over ambient noise levels. These levels can be difficult or impossible to achieve when isolation is poor. Similarly, noise intrusion to others in a house, or neighbors in an apartment may be a significant issue.

In many cases, when available, for noise reasons, a basement may be the best location for a home studio. It economically offers a high degree of isolation from the neighboring external environment, if not always from the upstairs house. A freestanding garage separated from a main house may not solve noise problems with neighbors, but would lower noise levels in the house.

As in any structure, the windows in a home are an acoustical weak link. External sound can easily penetrate most home windows, and likewise sound from within can emerge to annoy neighbors. Many kinds of windows can be covered with a device that is simple to construct. Much like a hurricane shutter, it shields the window glass, in this case, from noise. Figure 19-1 shows the design. A cover is constructed of several layers of 3/4-in particle board; for example, four layers will yield a transmission loss of about 40 dB at 500 Hz; the TL curve has an upward slope with respect to frequency. A wood frame is placed around the outer window sill; it is secured with adhesive. The cover is secured to the frame with carriage bolts; this makes the cover easily removable when not needed. Alternatively, the cover can be nailed to the frame. Perimeter seals are pressed into place. The cover can be placed on either the exterior or interior side of the window.

Despite the best efforts, a home studio will almost certainly have noise intrusion problems. It should be expected that external noise will sometimes ruin an otherwise satisfactorily recorded take. The cost of absolutely preventing this would far exceed the frustration of occasionally ruined takes. The more serious issue is the potential complaints from neighbors.

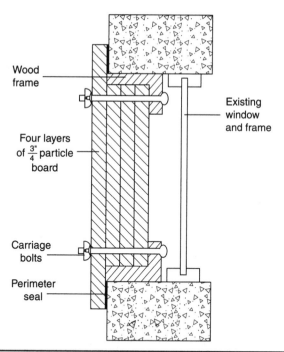

FIGURE 19-1 Noise intrusion through a window can be reduced by covering the window with a multilayer wood cover, sealed at the periphery.

Studio Design Budget

Most home studios are built on a strict budget. Although prefabricated acoustical modules greatly simplify construction and produce excellent acoustical results, their cost is beyond the reach of many home-studio budgets. Alternatively, modules can be constructed from simple building materials at low cost. These modules take time and effort to build, may lack the finished appearance of commercially manufactured units, and may not be acoustically ideal, but they will certainly be functional. Therefore, in this home studio design example, the three types of acoustical treatment (broadband absorption, bass traps, and diffusers) are designed to be built from scratch.

Studio Treatment

In an untreated control room or listening room, at the mixing or listening position, the desirable direct sound from the loudspeakers is degraded by early reflections from walls, floor, and ceiling. These reflections arrive at the listener at slightly different times, creating acoustical comb filters that degrade the frequency response of the direct sound. This is contrary to the fact that in any critical playback environment it is essential to have an accurate sound field for mixing and listening. To overcome this problem, this studio will use the concept of a reflection-free zone (RFZ). In an RFZ room design, these early reflections are eliminated or attenuated; this improves the accuracy of the stereo image, the breadth of the sound stage, and the general quality of the sound. The RFZ concept is discussed in Chap. 15.

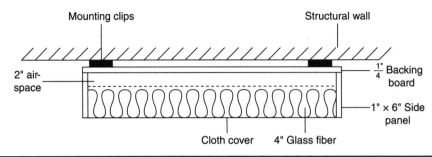

FIGURE 19-2   A wall-mounted absorptive panel.

Figure 19-2 shows an example of an absorption panel that can help create a reflection-free zone in this home studio. The panel is constructed with 1 × 6-in framing with a 1/4-in backing panel. The face of the unit measures 1 ft, 6 in × 4 ft. The interior contains a 4-in glass-fiber sheet or batt that is held in place with open wire mesh; the glass fiber is placed at the front of the frame with a 2-in airspace behind. The front of the panel is covered by an open-weave fabric. The panel can be attached to the structural wall with metal clips. The panel absorbs broadband sound; this is optimized by spacing the glass fiber away from the wall. Panels can be located on walls and ceiling between the loudspeakers and the mixing/listening position at locations where sound from the loudspeakers would otherwise reflect and then travel to the listening position. These locations can be determined by sitting at the listening position while an assistant moves a mirror across likely reflective surfaces. When the loudspeakers can be seen in the mirror, this identifies positions that should be covered by panels.

As noted above, small rooms are generally subject to low-frequency reverberation problems resulting from the modal resonances related to the room dimensions. In addition, in some rooms, most available absorption is at mid and high frequencies, so low frequencies are relatively unabsorbed and thus more prominent. This problem of "boomy" sound in small rooms is well known. Because of the long wavelength of low-frequency sound, effective absorbers require much space. It is wise to utilize the corners of a room because it is space that is often unused. More importantly, all modes terminate in corners, making this location most effective for bass absorption. Bass traps are discussed in Chap. 3.

The problem of boomy sound stems from lack of low-frequency absorption. This is often the case in rooms with walls constructed of masonry such as concrete blocks and brick. On the other hand, rooms constructed with gypsum-board stud walls can be free of this problem because walls of such construction are good natural low-frequency absorbers. Many homes are constructed with drywall partitions; thus bass traps might not be needed. This diaphragmatic absorption is quite effective in the 70- to 250-Hz region. The prudent course is to construct the room without additional low-frequency absorption, then measure and listen to the room. If the reverberation time is consistent across the audio spectrum, then the room has adequate low-frequency absorption. If the reverberation time at low frequencies is longer than at high frequencies, then bass traps or other low-frequency absorbers may be needed. The question of drywall low-frequency absorption is considered in Chap. 16.

The bass traps suggested for this studio are trapezoidal units that are designed for corner mounting, as shown in Fig. 19-3. The face of the unit measures 2 × 2 ft.

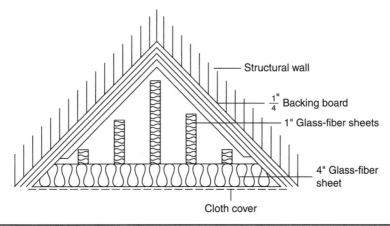

Structural wall

$\frac{1}{4}$" Backing board

1" Glass-fiber sheets

4" Glass-fiber sheet

Cloth cover

FIGURE 19-3   A corner-mounted bass trap module.

These units use internal glass-fiber sheets to provide absorption. Their relatively deep depth in the corner increases absorption at low frequencies. If desired, the cavities between the glass-fiber sheets can be loosely stuffed with glass-fiber batt.

Diffusion can be obtained from surface irregularities, polycylindricals, and other geometrical approaches. For example, excellent diffusion can be obtained from a series of parallel wells related to quadratic-residue number theory (Schroeder, 1975). Because the number theory is simple to implement, efficient diffusing surfaces based on the theory can be constructed by a carpenter. Modules using different well depths are described in Chap. 4.

In this studio design, a different implementation is used. The diffusing surface consists of an array of protrusions as shown in Fig. 19-4. The face of the unit measures 2×2 ft. This configuration provides omnidirectional diffusion of first-order and other reflections. Even if number theory is not used in the design, any irregular topographic surface will provide diffusion. However, if low-frequency diffusion is desired, the surface projections will have to be relatively large. When these modules are mounted on the rear wall, they will help create a diffuse field in the room. Modules can be mounted to the wall with metal clips. Diffusion is discussed in Chap. 4.

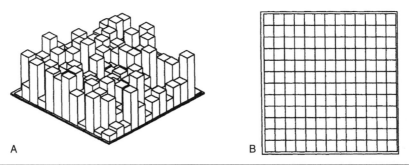

A   B

FIGURE 19-4   A wall-mounted diffusive module. (A) Isometric view. (B) Plan view.

Studio Design: Initial Treatment

Figure 19-5 shows the design for the proposed studio using low-cost, minimal treatment. To minimize costs, this initial treatment represents the very minimum treatment for the effects of early reflections, the minimum of bass absorption, and the minimum of rear-wall diffusion. Much potential improvement remains. In many ways, this design is similar to that of a home listening room. We begin by studying the properties of the room for playback; later, we will explore whether this same room can be used successfully for recording.

Two absorptive panels (1) are applied to each side wall. There will be some improvement in sound quality because many of the side-wall reflections are treated. However, the panels cover a limited surface area, so not all of the early reflections will be intercepted. The listening position will be somewhat troubled by early reflections, because reflections from the front wall between the loudspeakers are left untreated, as are reflections from the floor and ceiling. Two bass traps (2) are mounted in each of the two rear corners. They would help reduce boominess but may not completely eliminate it.

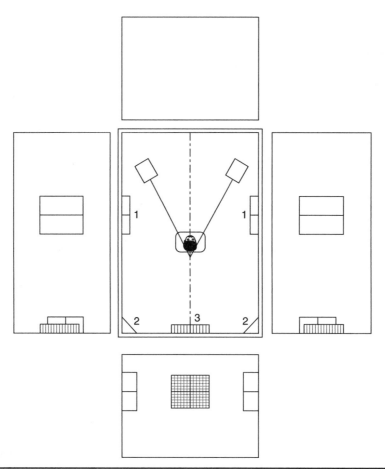

Figure 19-5 Home studio: Initial treatment.

Four diffusing units (3) on the rear wall intercept only a small fraction of the sound energy falling on the rear wall. These four units will not provide a truly diffuse field, but their presence will be noted as a modest improvement in the sound field.

Studio Design: Intermediate Treatment

Figure 19-6 shows an additional layer of room treatment. It represents the next logical progression toward a higher-cost comprehensive room design. In this design, the number of absorptive units (1) on each side wall is the same as in the earlier "initial treatment" design, but there are now four absorptive panels on the ceiling. This minimizes unwanted early reflections from the ceiling. In addition, the floor reflections could be reduced materially with a 6 × 6-ft rug. In addition, there are now four bass traps (2) in each of the two rear corners. If the ceiling height was limited to 8 ft, and the modules were used, this would be the maximum possible number in a corner. The diffusive panels (3) are left unchanged.

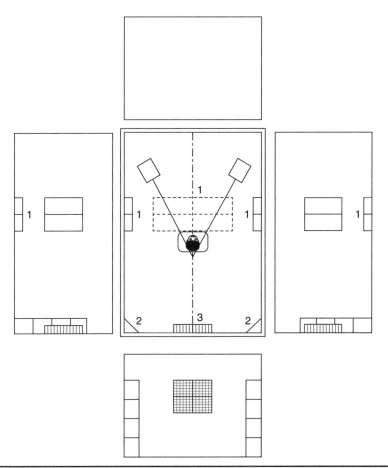

Figure 19-6 Home studio: Intermediate treatment.

Studio Design: Comprehensive Treatment

A comprehensive design is shown in Fig. 19-7. This more costly design essentially increases the number of all of the treatment units to the required level. Three absorptive panels (1) are now placed on each side wall, six on the ceiling and four on the front wall between the loudspeakers. Four bass traps (2) are now in each of the four corners of the room. Another option is to build a gypsum-board closet in a corner; its lightweight wall construction would let the closet double as a bass trap. Eight of the diffusive panels (3) are now mounted on the rear wall. This diffusion will help provide improved definition of the stereo image, accurately perceived depth of sound stage, and a feeling of being enveloped by the sound.

Overall, with this additional treatment, potentially troublesome acoustical problems should be eliminated. The room should provide neutral acoustics so that a mix done under these acoustical conditions should be transferable to other listening environments.

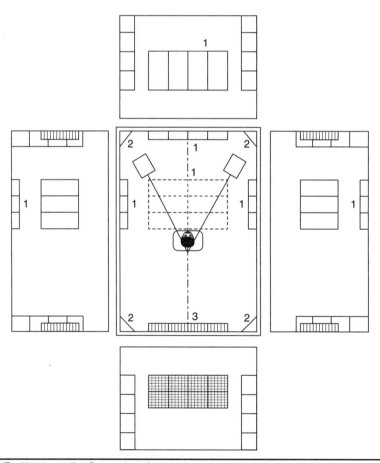

Figure 19-7   Home studio: Comprehensive treatment.

Recording in the Studio

The designs of Figs. 19-5, 19-6, and 19-7 are for mixing or listening with loudspeakers as the source of sound; these room designs essentially replicate a conventional RFZ control room. However, to complete this home studio design, the same room must also be used as a recording studio for music performance. In this case, the source of sound is a musician. Two questions arise: Where should musicians be positioned and where should the microphones be placed?

If the musicians are placed between the loudspeakers and the microphone is placed at the listening sweet spot, the recording would be quite free of early reflections. The side wall, ceiling, and floor absorbers were placed according to loudspeaker directionality. With a broader source, such as one or more musicians located between the two loudspeakers, the early sound would not be treated completely by these absorbers, but it would be partially treated.

One might ask whether it is desirable to record musicians without early reflections. Eliminating the early reflections from the walls, ceiling, and floor of the mixing/listening room has been for the purpose of keeping the music from the loudspeakers free of them. This helps ensure that sound from the loudspeakers is completely neutral at the listening position. The chances are good that music coming over the loudspeakers has its own collection of early reflections as part of its original capture. Therefore, any effort to record without early reflections could be considered specious, that is, having a false aspect of genuineness.

Nevertheless, the room shown in Fig. 19-7 would provide a good recording venue. The listening chair could be pushed to one side and this main space of the room used by musicians. The absorbers on side walls, front wall, and ceiling will give a good reverberatory condition. The bass traps in the corners will give a "tight" bass sound. The sense of openness contributed by the large panel of diffusers will be greatly appreciated by the musicians. In short, all the elements are present for a good recording studio, even though they have all been placed with playback in mind. Also, the acoustical asymmetry of the room as it has been treated will give a small degree of variability during recording; placing musicians in different locations in the room will change the nature of the sound recorded from them.

The studio of Fig. 19-7 (or, to a lesser degree, its early stages of Figs. 19-5 and 19-6) can be considered a good mixing/listening room as well as a good home recording studio. The acoustical elements that have been added are quite basic in their function, and are highly adaptable to space available in a home.

Garage Studio

A freestanding 2-car garage would provide a degree of isolation from the main house as well as sufficient floor space for a home studio. A typical garage might measure 24 × 24 ft, yielding a floor area of 576 ft². As shown in Fig. 19-8, a partition (1) could be placed diagonally across one corner, yielding a control room of about 176 ft² and a studio of 400 ft². These are minimal room sizes, but are workable. (In the previous example in this chapter, a rectangular space is used, and the control room and studio are integrated into one room.) The partition should be a staggered-stud wall or a double wall with multiple layers of gypsum board on both sides. An observation window (2) should be double pane with each pane independently set into each wall. Microphone panels and other

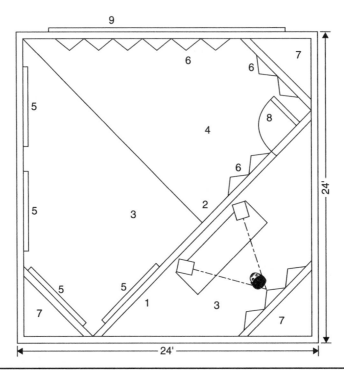

FIGURE 19-8 Floor plan of garage studio.

fixtures on the common wall should be staggered (not located back-to-back) and all openings and conduits should be sealed. Although it is probably not practical, and the benefit would be fairly minimal, the studio and control room could be further isolated by cutting the concrete slab to yield two independent slabs. The control room should be constructed as a room within the garage, with separate walls, floating floor, and ceiling. Consideration should be given to heating and ventilating each room; clearly, there must not be any common ductwork between the studio and control room.

Both rooms should be relatively dead with reverberation times below 0.5 sec. Carpet (3) and pad can be laid on a portion of the concrete slab, and tile (4) or parquet on another portion. A mixture of absorptive panels (5), reflecting panels and diffusers (6) should be placed on the walls and ceiling. The control room has axial symmetry with the console facing the diagonal wall; in such a small space, the monitor speakers should be placed in a near-field configuration. A combination resonator/diffuser (7) can be placed in the rear corner as well as in corners of the studio. Any single doors (8) should be solid core, and tightly weather stripped. Clearly, any open louvers and glass windows should be sealed, using the wood covers as described above.

For greater recording diversity, even in such a small studio, one end of the studio should be more absorptive, while the other end is more live. For example, generally, drums will be recorded in the dead end of the studio. Isolation in drum tracks can be enhanced by suspending a thick absorptive panel over the drums. Moreover, a bass trap can be located in the space above the absorbing panel; a bass trap can be constructed by hanging rows of glass-fiber boards vertically. Conversely, an acoustic guitar would be recorded in the live end of the studio. The walls there are reflective and diffusive, and the floor is left uncovered.

On the downside, a garage might provide little sound isolation. It might be of lighter construction than the main house, and may be situated close to a neighbor's house. Isolation can be improved by adding a layer of cement board to the exterior and a double drywall layer on the inside; the latter should be mounted on isolators or independent framing. All seams should be taped or sealed with nonhardening acoustical sealant. Because garage doors use lightweight construction, they offer little insulation. A new wall should be constructed behind the garage door (9), leaving the garage door as a facade.

CHAPTER 20

Home Listening and Media Room

M ost homes have a central "living room" that is used for a variety of leisure activities; some of the technology-related activities include listening to music, watching television, surfing the Internet, video gaming, and video chatting. Equipment usually includes a stereo, flat-screen TV, game console, and various kinds of computers. These rooms are rarely designed and treated with acoustical performance in mind. But, with fairly minimal design changes, the acoustical aspects of the activities there can be greatly improved.

Home listening and media rooms present unique acoustical challenges, but also offer creative opportunities. Unless you are building a new home or a new addition to a home, the size and geometry of your listening room is probably predetermined. Likewise this predetermines the room's low-frequency response. However, within aesthetic bounds, there is great freedom in placing bass traps, absorbers, and diffusers to optimize the room's response. In addition, there may be options for the placement of the loudspeakers and listeners. All of these decisions represent variables that can greatly affect the resulting response at the principal listening position.

Several examples of listening room designs are presented in this chapter. In particular, three acoustical treatments, progressing from simple to sophisticated, are given for one room geometry. As with any small-room design, priority is given to room modes and the effect of reflections on the quality of sound. More specialized designs of dedicated home theaters are presented in the following chapter.

Low-Frequency Response

At the low-frequency end of the spectrum, audio wavelengths are long. For example, at 100 Hz the audio signal wavelength is 1130 ft/sec divided by 100 or 11.3 ft per cycle. The wave nature of these long-wavelength sounds creates a modal response; this is the source of a large portion of the problems normally encountered in small-room design. At the mid- or high-frequency end of the spectrum, the wavelengths of sound are relatively short. For example, at 300 Hz, the wavelength of sound is 1130/300 Hz, or 3.8 ft per cycle. It is acceptable to think of short-wavelength sounds as rays reflecting in the room, with relatively fewer problems. Very generally, 300 Hz is often considered to be the cutoff frequency between these two ways of considering sound. Because the transition between the two regions is a gradual one, the use of 300 Hz is only a convenience. The larger the room, the lower the transition frequency.

At low frequencies, the room dimensions are of the same order as the sound being delivered by the loudspeakers. For example, your subwoofer probably delivers significant energy at 50 Hz, at which the wavelength is 23 ft long. So, at 100 Hz the wavelength of sound may be about equal to the width of your room, and at 50 Hz it may equal to the length. Instead of a space uniformly packed with sound rays (above 300 Hz), the room resonates at various audible frequencies (below 300 Hz). To obtain satisfactory bass response at the listening position, the room modes that are specific to that room must be considered. Chapter 5 contains more information on the topic of modal response.

Modes in Typical Rooms

The following room sizes will be examined:

Small room:	19 ft, 6 in × 10 ft, 6 in × 8 ft, 0 in
Medium room:	24 ft, 0 in × 14 ft, 6 in × 9 ft, 0 in
Large room:	27 ft, 9 in × 17 ft, 0 in × 9 ft, 6 in

There are larger and smaller rooms that are used for listening environments, but there is much to learn from these three. Table 20-1 lists the frequencies of the axial modes

	Length $L = 19.5$ ft $f_1 = 565/L$	Width $W = 10.5$ ft $f_1 = 565/W$	Height $H = 8$ ft $f_1 = 565/H$	Arranged in Ascending Order	Difference
f_1	29.0 Hz	53.8 Hz	70.6 Hz	29.0 Hz	24.8 Hz
f_2	57.5	107.6	141.3	53.8	3.7
f_3	86.9	161.4	211.9	57.5	13.1
f_4	115.9	215.2	282.5	70.6	16.3
f_5	144.9	269.0	353.1	86.9	20.7
f_6	173.8	322.9		107.6	8.3
f_7	202.8			115.9	25.4
f_8	231.8			141.3	3.6
f_9	260.8			144.9	16.5
f_{10}	289.7			161.4	12.4
f_{11}	318.7			173.8	29.0
				202.8	9.1
				211.9	3.3
				215.2	16.6
				231.8	29.0
				260.8	8.2
				269.0	13.5
				282.5	7.2
				289.7	29.2
				318.9	
				Mean difference: 15.26	
				Standard deviation: 8.9	

TABLE 20-1 Axial Modes in a Small-Sized Home Listening Room (Room Dimensions: 19 ft, 6 in × 10 ft, 6 in × 8 ft)

below 300 Hz for the small room (listed above). All of the entries in this table are active resonances at specific frequencies that will affect low-frequency music components.

The lowest frequency of modal resonance in this small room is 29 Hz, which is the first-order mode associated with the length of the room. Sound energy at 29 Hz will reflect back and forth between the front-end and back-end walls. The sound pressure over the surface of the end walls will be high, and there will be a vertical null plane at the center of the room (it would be instructive to inspect Fig. 16-2 at this point). If this null plane coincides with a listener's head at the sweet spot, he or she would theoretically perceive no 29-Hz component of the signal. At the rear or front walls, however, the energy level at 29 Hz would be at a maximum.

Table 20-1 shows not only the first-order frequency of 29.0 Hz, but also 53.8 Hz (the first-order mode between the two side walls) and 70.6 Hz (the first-order mode of the resonance between the floor and the ceiling). Moreover, the modes at integral multiples (second-, third-, fourth-order, etc.) are similar in amplitude to the modes at first-order frequencies.

Mode Spacing

Ideally a room would exhibit a smooth response curve throughout the audio range. In practice, this is never the case, especially at low frequencies. The acoustical response curve of the small room of Table 20-1 below 300 Hz can be imagined by plotting the 19 narrow-response curves on a linear frequency scale. All 19 have approximately the same amplitude. It would be fair to assume a bandwidth of 4 Hz for each resonance, measured between the −3-dB points. A glance at such a plot would emphasize the need for uniform distribution of resonances, which is determined by the spacing between adjacent resonances. However, the problem is not quite this simple. The various modes exist at different frequencies and also at different areas in the room; as a result, at any given place in the room, the modal peaks or nulls are effectively spaced even further apart than their numerical frequency spacing indicates.

Evenness of distribution of resonances along the frequency scale is determined by the relative proportions of the room dimensions. In a cubic room, the length, width, and height first-order resonances would be identical, each multiple resonance would have triple energy, and the spacing between modal resonances would be great. Careful proportioning of length, width, and height serves to distribute the resonances more uniformly (see Table 5-4).

In Table 20-1 the modal frequencies below 300 Hz are arranged in ascending order, and from these the spacing of adjacent modes is determined by calculating the difference in frequency between adjacent modes. The spread of differences is from 3.3 Hz to 29.2 Hz, but the average difference is 15 Hz. Sixty-seven percent of the differences are within 8.9 Hz of 15 Hz. The latter is called the standard deviation. Studies of audible colorations of speech have shown that 15-Hz separation of adjacent modes is usually acceptable, but spacings greater than 25 Hz and zero spacing (a coincidence) can both cause colorations.

Low-Frequency Peaks and Nulls

Table 20-2 lists low-frequency resonant frequencies in a medium-sized room and Table 20-3 lists the resonance frequencies in a large-sized room (both listed above). As noted, it is not accurate to imagine the three columns of frequencies acting at any single point.

	Length L = 24 ft $f_1 = 565/L$	Width W = 14.5 ft $f_1 = 565/W$	Height H = 9 ft $f_1 = 565/H$	Arranged in Ascending Order	Difference
f_1	23.5 Hz	39.0 Hz	62.8 Hz	23.5 Hz	15.5 Hz
f_2	47.1	77.9	125.6	39.0	8.1
f_3	70.6	116.9	188.3	47.1	15.7
f_4	94.2	155.9	251.1	62.8	8.0
f_5	117.7	194.8	313.9	70.6	7.3
f_6	141.3	233.8		77.9	16.3
f_7	164.8	272.8		94.2	22.7
f_8	188.3	311.7		116.9	0.8
f_9	211.9			117.7	7.9
f_{10}	235.4			125.6	15.7
f_{11}	259.0			141.3	14.6
f_{12}	282.5			155.9	8.9
f_{13}	306.0			164.8	23.5
				188.3	0.0
				188.3	6.5
				194.8	17.1
				211.9	21.9
				233.8	1.6
				235.4	15.7
				251.1	7.9
				259.0	13.8
				272.8	9.7
				282.5	23.5
				306.0	
				Mean difference: 12.29	
				Standard deviation: 7.05	

TABLE 20-2 Axial Modes in a Medium-Sized Home Listening Room (Room Dimensions: 24 ft × 14 ft, 6 in × 9 ft)

Instead, each of those modal frequencies affects the listening room's response differently throughout its space.

Figure 20-1 shows the first three resonances of only the length and width columns of the medium-sized room. Taken from Table 20-2, the sound pressure lines for 23.5 and 47.1 Hz are shown at the top of Fig. 20-1, and 70.6 Hz is shown at the bottom of the figure. A sound-pressure peak is always present at the ends of the room and between nulls. Referring to the room width column of Table 20-2, the frequencies of 39.0, 77.9, and 116.9 are selected and sketched onto Fig. 20-1. The nulls for these six resonances are indicated as broken lines on the floor plan. These lines are really the bottom edge of null planes reaching to the ceiling (see Fig. 16-2).

	Length L = 27.7 ft $f_1 = 565/L$	Width W = 17 ft $f_1 = 565/W$	Height H = 9.5 ft $f_1 = 565/H$	Arranged in Ascending Order	Difference
f_1	20.4 Hz	33.2 Hz	59.5 Hz	20.4 Hz	12.8 Hz
f_2	40.8	66.5	118.9	33.2	7.6
f_3	61.3	99.7	178.4	40.8	18.7
f_4	81.7	132.9	237.9	59.5	1.8
f_5	102.1	166.2	297.4	61.3	5.2
f_6	122.5	199.4		66.5	15.2
f_7	142.9	232.6		81.7	18.0
f_8	163.4	265.9		99.7	2.4
f_9	183.8	299.1		102.1	16.8
f_{10}	204.2			118.9	3.6
f_{11}	224.6			122.5	10.4
f_{12}	245.0			132.9	10.0
f_{13}	265.4			142.9	20.5
f_{14}	285.9			163.4	2.8
f_{15}	306.3			166.2	12.2
				178.4	5.4
				183.8	15.6
				199.4	4.8
				204.2	20.4
				224.6	8.0
				232.6	4.9
				237.5	7.5
				245.0	20.4
				265.4	0.5
				265.9	20.0
				285.9	11.5
				297.4	1.7
				299.1	
				Mean difference: 9.25	
				Standard deviation: 8.67	

TABLE 20-3 Axial Modes in a Large-Sized Home Listening Room (Room Dimensions: 27 ft, 9 in × 17 ft × 9 ft, 6 in)

Only six of the 23 room resonances below 300 Hz from Table 20-2 are included in Fig. 20-1. If all 23 were plotted, null-plane edges would cover the floor and walls, and peak planes would be just as plentiful. There is no avoiding them, nor do we want to avoid them, but it would be beneficial if the nulls were not so deep and the peaks were not so high. That is what low-frequency absorption can accomplish.

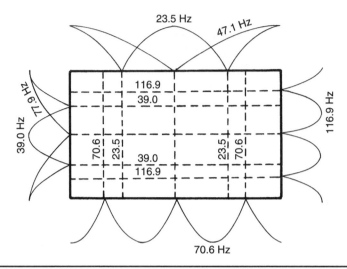

FIGURE 20-1   Some of the null-plane edges in a medium-sized home listening room.

Imagine sitting in the room of Fig. 20-1 with all peaks and nulls in place while a musical selection is being played. Different modes are excited as the musical notes go up and down the scale and chords (and their overtones) are played. The room's dimensions thus reinforce and cancel sound waves, resulting in the room's three-dimensional frequency response, a response that affects the response of the music you hear, depending on the placement of the loudspeakers, and the listeners. In fact, it would be advantageous to plot the room response on the floor plan and find the optimal layout of the loudspeakers and listeners.

Proper low-frequency acoustical treatment of the room will tend to smooth out the modal peak-null differences, and will minimize the variation of response with position in the room. A room treated, for example, only with a foam material of 1-in thickness will not affect the modes because it offers practically no absorption in the low-frequency region. Low-frequency absorption requires special techniques, devices, and effort. At higher frequencies (above the bass region), the modal peak-null spacing is so dense that the response is relatively smooth. At higher frequencies, room treatment is used to address other issues such as reflection and diffusion.

Effects of Room Size

Tables 20-1, 20-2, and 20-3, for a small-, medium-, and large-sized room, respectively, also give us the opportunity to study the effect of room size on mode spacing, as summarized in Table 20-4. All statistics are for modes lower than 300 Hz.

Table	Size	Mean Mode Spacing	Standard Deviation
20-1	Small	15.26 Hz	8.9 Hz
20-2	Medium	12.29	7.05
20-3	Large	9.25	8.67

TABLE 20-4   Comparison of Mode Statistics for Three Rooms

The dense 9-Hz mean mode spacing in the large room underlines the acoustical value of large venues. The larger the room, the lower the frequency placement of the lowest modes. In very large rooms, the lowest modes, and the relative irregularity of the resulting frequency response, are pushed below the frequency-response region of interest. The constancy of the standard deviation suggests that similar factors of spacing are at work in all three sizes of rooms.

Loudspeaker Positioning

Sound from a loudspeaker interacts with any reflecting surfaces near it. For example, placing a loudspeaker symmetrically in a tricorner adds a 9-dB low-frequency peak to its response. Change this location from a tricorner to a spot on the floor close to a wall, and the low-frequency peak is 6 dB. Place it close to an isolated wall and the peak is only 3 dB. If your system needs the low-frequency peak, these loudspeaker placements will give it. The peak cannot be defeated if the loudspeaker is near a surface, but it can be minimized by placing the loudspeaker at different distances from each of the three reflecting surfaces.

True dipole loudspeakers, such as the electrostatic type (which have a tall, vibrating membrane), have a strong rear radiation component that must be controlled to prevent another source of early reflections. The more common dynamic loudspeakers radiate rearwards weakly, chiefly by the diffraction from corners of the cabinet.

Acoustical Treatment for the Listening Room

When treating a listening room, one must consider the time of arrival of sound waves. Because of room reflections, an original sound wave will arrive repeatedly and delayed in time because of the varying path lengths of the reflections. This causes irregularities in the signal's frequency response. To overcome this, the early reflections should be absorbed or diffused at the reflecting surfaces. It is also important to control the low-frequency modal resonances of the room. To do this, bass traps or similar absorbers should be placed in the corners of the room. Once the effect of placing low-frequency absorbers in two corners is heard, the desirability of doing the same in the other two corners will undoubtedly be evident. For example, TubeTraps (manufactured by Acoustic Sciences Inc.) or Modex Corner absorbers (manufactured by RPG Diffusor Systems, Inc.) (see Chap. 16) may be used. Alternatively, bass traps can be custom built (see Chap. 19).

The room should not be acoustically too dead or too bright. The typical living room reverberation time of about 0.5 sec should be close to what is needed. If there is too much overstuffed furniture, the room might be too dead; some absorptive furniture may have to be removed. If the room seems too bright, fabric-covered glass-fiber panels (see Chap. 19) could be introduced to achieve the best balance.

Because it is such an important element, diffusion has been emphasized in many designs in this book. Diffusers mounted on the rear wall of the listening room are encouraged (see Chap. 19). Also, as we shall see later in this chapter, in dead rooms, diffusers can be preferable to absorption for controlling early reflections, because diffusion does not decrease desirable signal energy.

Identification of Early Reflections

Figure 20-2 identifies significant reflective surfaces in the front end of a rectangular listening room. In addition to the direct sound from the loudspeaker to the listener, the loudspeaker's signal is reflected from these surfaces, and the reflections interact with the direct sound and with each other, to create comb-filter distortion, which degrades the sound the listener hears.

The first-order (single-bounce) reflections are classified as early reflections because they are the first reflections to arrive after the direct sound. The program quality perceived by the listener in a typical listening room is greatly affected by these early reflections. Assuming reasonable care in controlling the effect of the low-frequency modal resonances, and assuming source, amplifiers, and loudspeakers of reasonable quality, the early reflections are undoubtedly the major determinant of program quality.

One way to reduce the effect of early reflections is to cover the reflective area with absorbing material. It is easy to find these reflective surfaces. For example, consider the area on the floor (Fig. 20-2) where sound from the tweeters of the two loudspeakers is reflected. To identify the area, to lay a mirror on the floor and have an assistant move it until you (the listener) can see a tweeter. A single rug (1) can cover areas from both stereo tweeters, and the floor reflection is thus controlled.

Similarly the positions at which absorbent should be placed to control the left and right side-wall reflection (2) can be found by using a mirror. The location for the panels to absorb the ceiling reflections (3) is more difficult to find with a mirror, but the

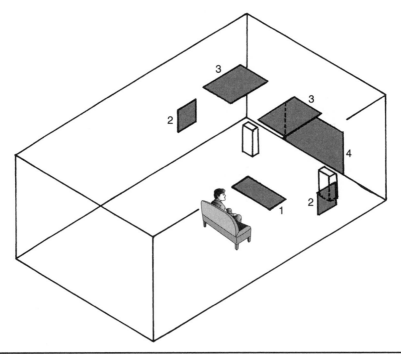

Figure 20-2 An isometric drawing of a home listening room showing the location of spot absorbers.

experiments to find areas (1) and (2) probably suggest a good pair of spots, and success can be ensured by using an absorbent of generous size. The absorbent behind the loud-speakers (4) can be logically located by knowing that the sound source is the edge of the loudspeaker cabinets.

Psychoacoustical Effects of Reflections

Before controlling the early sound reflections using the method described above, it is urged that the condensed results of the research of Floyd Toole and Sean Olive (Olive and Toole, 1989) on lateral reflections be reviewed (see Chap. 11). These reflections affect a listener's impression of spaciousness, in which the listener feels immersed in the music. Shifting and spreading of the auditory image are also a function of reflection amplitude. In other words, one does not want to eliminate lateral reflections, but rather to control them to yield a degree of spaciousness. With some thought, and some experimentation, it is possible to adjust lateral reflections and achieve a sense of appropriate spaciousness in even a small listening room.

Examples of Listening Room Treatment

This section on treating a listening room will have many similarities to the discussion of home studios in Chap. 19. In that chapter, designs for an absorptive panel, bass trap, and diffusive panel were introduced as cost-effective units that are attractive for budget-limited designs. The three listening-room designs to follow describe a single room with progressive additions to improve the performance of the room. Each design is acoustically optimized for that budget level. Both scratch-built and commercially manufactured units will be discussed. The room dimension ratio is 1.0:1.4:1.9.

Listening Room Design: Initial Treatment

The home listening room is shown in Fig. 20-3; this basic design demonstrates how treatment can be done on a budget. The side-wall reflections are treated on a minimum basis with a single absorption panel (1) (see Chap. 19) placed on each side wall to provide broadband absorption and reduce reflected energy. Alternatively, an Abffuser panel (manufactured by RPG Diffusor Systems, Inc.) can be placed on each side wall.

The ceiling reflection can be controlled with a pair of similar absorption panels (2). Alternatively, a Nimbus ceiling cloud (manufactured by Primacoustic) can be used. These panels are thin fabric-covered absorbers spaced out from the ceiling to improve absorption. The panels use 2-in-thick, 6-lb/ft^3 glass fiber, and measure 2 × 4 ft. Because it is of secondary importance, there is no treatment of the wall behind the loudspeakers at this budget level.

Some control of low-frequency modes is offered by two trapezoidal bass traps (3) (see Chap. 19) located in the rear corners. Alternatively, two Modex Corner absorbers (manufactured by RPG Diffusor Systems, Inc.) can be used. This product is available with a peak absorption at 40, 63, or 80 Hz; the latter two approximately coincide with the frequency of the first ceiling mode (575/8 = 70.6 Hz) for 8-ft ceilings, which is a common ceiling height.

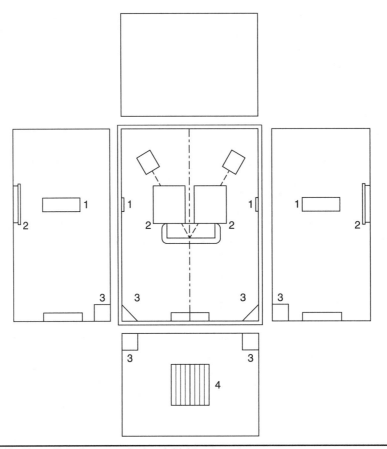

FIGURE 20-3 Home listening room design: Initial treatment.

Two diffuser panels (4) (see Chap. 19) are mounted on the rear wall. They return a broad spatial pattern of diffuse reflections to the listening position. Alternatively, QRD-734 diffusing units (manufactured by RPG Diffusor Systems, Inc.) could be mounted on the rear wall. These wooden units are based on prime 7 quadratic-residue theory, with a depth of 9 in. This room should perform well for noncritical listening; there are additional improvements reserved for the room designs that follow.

Listening Room Design: Intermediate Treatment

The home listening room is shown again in Fig. 20-4; this design is an upgrade of the basic listening room. This design demonstrates where additional treatment should be placed to most cost-effectively improve the room acoustics. In particular, this design doubles the area of the treatment on the side walls and rear wall. Two broadband absorption panels are now located on each side wall, and two more diffuser panels are added to the rear wall for a total of four. Two more bass traps are also added to the rear corners. The acoustics of this room will be quite suitable for most listening needs.

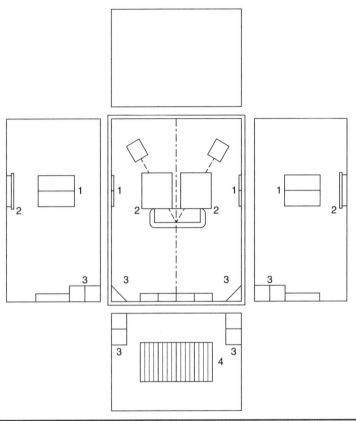

Figure 20-4 Home listening room design: Intermediate treatment.

Listening Room Design: Comprehensive Treatment

The home listening room is shown again in Fig. 20-5; this design provides acoustically optimal, albeit more expensive, treatment. The principal change in the listening room is treating the wall behind the loudspeakers. The front wall now has two diffuser panels (4) placed on the inside and two absorption panels (5) as the outside units. If the floor has a hard surface, it is also suggested (but not shown in the figure) that a rug be placed in front of the sofa to control reflections from the floor.

The acoustics of this listening room should be of excellent quality. The imaging should be clear and definite over the entire sweet spot, and the sweet spot should be much larger than in most rooms. Listeners should experience a high degree of envelopment in the music and, with controlled early reflections, frequency response at the listening position should be excellent.

Background Noise

In an existing home, most of the typical noise producers are unavoidably present. The central HVAC system is probably of the small-duct, high-velocity air type and will

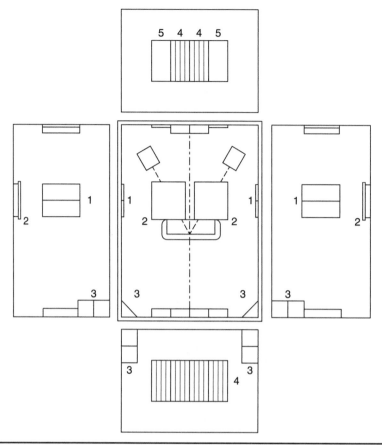

Figure 20-5   Home listening room design: Comprehensive treatment.

generate air-turbulence noise. The water pipes are probably fastened solidly to the structure, so plumbing noise will unfortunately be efficiently radiated into the rooms. The noise of daily living is cheerfully generated by the home's occupants. In other words, there is very little that can be done to reduce background noise. The lucky person might possibly be able to choose a room at some distance from the highest levels of household clatter, and install a solid-core door with gasketed seals. This will help reduce noise intrusion into the listening room, and conversely reduce noise intrusion to the rest of the house.

CHAPTER 21

Home Theater

N ever have the planets of playback equipment and media been as aligned for the home theater as they are today. Blu-ray disc players, surround receivers, surround loudspeaker systems, powered subwoofers, flat-panel HDTV screens, and projectors offer a scope and quality unimagined in the past. Sound and picture quality available in the consumer's home compares favorably with the best in commercial theaters.

A knowledgeable home-theater owner knows that the acoustical path between the loudspeaker diaphragms and his or her ear is as important as the quality of audio hardware such as amplifiers and loudspeakers. However, this acoustical path is less tangible and less available for correction and adjustment than the hardware. A first-class home theater needs to have excellent sound quality, but that is impossible without sophisticated acoustical treatment of the space housing the loudspeakers and the listener.

The home-theater room design presented here emphasizes elements that promote sound diffusion. The discovery in the 1960s of the application of number theory to practical sound-diffusing devices has increased the quality of professional sound recording and playback, as well as playback in home theaters. Proper diffusion helps promote a broad and expansive sound field, clarity of sound, and a sense of being thoroughly immersed in the music and sound effects.

Locating the Home Theater

Ideally, a home theater should be placed in a dedicated room. If the goal is to truly simulate the qualities of a motion-picture theater, a larger room volume is required than is otherwise needed for simple high-fidelity listening. A large, dedicated space will be assumed here. Possible compromises will be considered at the end of the chapter. The room can be rectangular, and if care is taken to eliminate flutter echoes, the parallel surfaces should pose no other particular problems. In other words, splaying of walls is probably not needed.

Whether the home-theater space considered here will be new construction or an adaptation of an existing space, there is need to consider room proportions. First-class acoustics is the goal (as well as first-class visuals) and the distribution of room resonant frequencies is the logical first consideration in achieving good acoustics. Small rooms are acoustically more difficult than larger ones because of the dominance of modal resonant frequencies that are too few and far between. Special attention is required to control them.

Home Theater Plan

Bass response is largely dictated by room dimension ratios. When designing a new room, it will be possible to choose room dimension ratios to minimize the effect of bass mode nulls in the listening area. However, in most cases, an existing room is used; therefore, its dimensions are predetermined. The task then is to calculate and plot room modes, and determine what measures must be taken to provide consistent bass response in the listening area. It may also be expedient to choose an optimal listening position based on the mode analysis. The floor plan of Fig. 21-1 shows a room measuring 8 ft × 12 ft, 10 in × 18 ft, 7 in, which follows the ratio of 1.00:1.60:2.33 (see Table 5-4).

A study of the axial modal frequencies below 300 Hz in this room is given in Table 21-1. The difference column reveals a rather coarse (because of the relatively small room size) field of axial modes but they are well distributed. The average spacing is almost 15 Hz, and there are numerous spacings in the 20-Hz region. It is evident that the axial modes are spaced as well as can be expected for a room this size. Low-frequency absorption and much diffusion should make the room very acceptable for listening.

Figure 21-1 shows the placement of furniture and playback equipment. The two main loudspeakers are arranged in a normal configuration; the included angle is typically 60° or slightly more. If possible, in most cases, the loudspeakers should be spaced

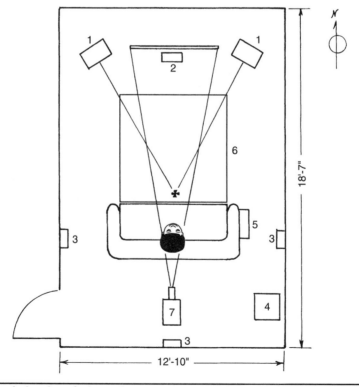

FIGURE 21-1   A plan view of a home theater.

	Length $L = 18.6$ ft $f_1 = 565/L$	Width $W = 12.8$ ft $f_1 = 565/W$	Height $H = 8$ ft $f_1 = 565/H$	Arranged in Ascending Order	Difference
f_1	30.4 Hz	44.1 Hz	70.6 Hz	30.4 Hz	13.7 Hz
f_2	60.7	88.3	141.3	44.1	16.6
f_3	91.1	132.4	211.9	60.7	9.9
f_4	121.5	176.6	282.5	70.6	17.7
f_5	151.9	220.7	353.1	88.3	2.8
f_6	182.3	264.8		91.1	30.4
f_7	212.6	309.0		121.5	10.9
f_8	243.0			132.4	8.9
f_9	273.4			141.3	10.6
f_{10}	303.8			151.9	24.7
				176.6	5.7
				182.3	29.6
				211.9	0.7
				212.6	8.1
				220.7	22.3
				243.0	21.8
				264.8	8.6
				273.4	9.1
				282.5	21.3
				303.8	
				Mean difference: 14.7	
				Standard deviation: 8.8	

TABLE 21-1   Axial Modes in a Home Theater (Room Dimension: 18 ft, 7 in × 12 ft, 10 in × 8 ft)

away from the wall by a foot or so; rear reflections can affect low-frequency fidelity. In this design, the sofa is placed somewhat behind the "sweet spot" to widen the sweet spot a bit. Maintaining reflections from the side walls will provide a wide area of good perception of the front soundstage, extending it over the entire sofa. Conversely, placing absorbers on the side walls will provide a more focused soundstage. These demands must be balanced. Diffusion in the back of the room will add spaciousness and envelopment. A good approach is to selectively add treatment a little at a time while listening to the effect of each addition. It is best to use any one treatment moderately; it is possible to over-treat a room.

The 10-ft distance to the television (a distance usually slightly shorter than the distance between the speakers) requires a relatively large television display for optimum viewing. Alternatively a video projector can be used. Placing a video projector in front of the viewers is rejected because sound reflections from the projector would affect the sound quality. However, when placed at the rear of the room, sound reflections from the video projector will actually contribute to the room ambience. A five-channel surround processor is needed, plus a left and right stereo loudspeaker (1), a center loudspeaker (2), surround loudspeakers (3), and a sub-woofer (4). The room may also include a control position (5). As an initial treatment element, if the floor is reflective, a 6 × 6-ft rug (6) might be placed in front of the sofa. The positions of the loudspeakers are predetermined except for the subwoofer. At the low frequencies that it radiates, directional effects are minimal. However, placement of the subwoofer in the room will greatly influence its bass response at the listening position. This is discussed below.

Early Reflections and Their Effects

Another floor plan, shown in Fig. 21-2, is introduced to show reflections that will affect the design of the home theater. To avoid confusion, some of the elements in the floor plan of Fig. 21-1 are omitted.

To simplify, four sound rays from the front left loudspeaker will be considered. Sound ray (A) from the left loudspeaker is reflected from the left side wall to the

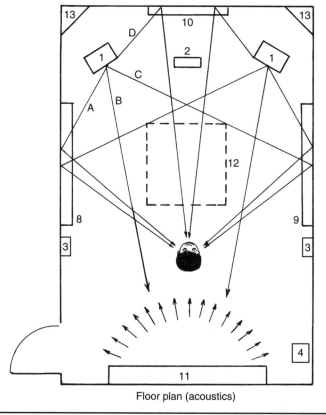

Floor plan (acoustics)

FIGURE 21-2   Plan view of a home theater showing surface reflections.

listener's ears. Sound ray (B) from the left loudspeaker travels to the rear of the room and will be discussed later. Sound ray (C) from the left loudspeaker strikes the right side wall and is reflected directly to the listener. Sound ray (D) from the left loudspeaker, generated at the corners of the cabinet by diffraction, travels from the corner of the cabinet to the front wall and is then reflected to the listener.

Sound rays (A), (C), and (D) all carry the same program material, but reach the listener at slightly different times because of the different reflection path lengths. These rays combine with each other and with the direct ray. When two signals add in phase, a peak is produced; when they add in phase opposition, they cancel, creating a null. A comb-filter response is formed; a nominally flat response is changed to one having alternating peaks and valleys through the spectrum. To minimize audibility of this distortion, the amplitude of the early reflections must be reduced (see Figs. 15-1 and 15-2).

Controlling Early Reflections

Our goal is to produce an essentially reflection-free zone (RFZ) at the listening position, so sound can be heard coming from the loudspeakers without the distortion caused by early reflections. This can be accomplished by mounting absorbing panels. First, panels are placed on the side walls at strategic points where they will intercept rays (A) and (C). Ray (D) will require another such panel on the front wall behind the loudspeakers (see Fig. 21-2).

Broadband absorbing panels using glass-fiber sheets (see Chap. 19) can be used for this application. The absorbing panels for the side and front walls (8, 9, and 10) are shown in Fig. 21-2. Alternatively, if the room is already highly absorptive, instead of absorptive panels, diffuser panels could be mounted in these locations. As another option, Abffusors (manufactured by RPG Diffusor Systems, Inc.) can be used; they function both as absorbers and quadratic-residue diffusers. In either case, the panels can be mounted flat on the wall with no airspace behind them. If spaced away from the wall, their absorption would be increased; sound can enter the panel from both the bottom and top sides. For example, a 4-in airspace could be used.

Reflections from the ceiling necessitate another absorber located on the ceiling between the loudspeakers and the listening position. A panel using the same design as the absorptive wall panels can be used. It is shown in the figure as (12). Alternatively, one or two Nimbus ceiling clouds (manufactured by Primacoustic) can be used. These panels are thin fabric-covered absorbers designed to be spaced out from the ceiling to improve absorption. To reduce the amplitude of the floor reflections a 6×6-ft rug, (6) in Fig. 21-1, is recommended. If properly placed and of the proper material, these absorbing (or diffusing) panels and rug should reduce the amplitudes of the early reflections and minimize the distortion they cause.

Ray (B) (Fig. 21-2) and its many counterparts from both loudspeakers travel directly to the rear of the room where they encounter an array of diffusers (11). A variety of diffuser designs could be used. For example, a diffusing surface consisting of protrusions can be used (see Chap. 19). This configuration provides omnidirectional diffusion. Even if number theory is not used in the design, other irregular topographic surfaces will suffice, albeit less efficiently. Alternatively, eight QRD-734 diffusers (manufactured by RPG Diffusor Systems, Inc.) could be used. When these panels are used, they should be oriented, so the four upper diffusers diffuse sound vertically, and the four lower ones diffuse sound horizontally. Many rays sweep past the listener and are reflected from room surfaces on their way to the rear of the room. Some of this energy strikes the diffusers, and some of the energy strikes the wall. Thus energy is returned to the listener either in specular or diffuse form.

Other Treatment Details

Figure 21-3 shows the east and west elevations of the room. These drawings illustrate the early reflections from the floor and ceiling and the absorbers for controlling them. As described above, the side-wall absorbers (8 and 9) and the absorber behind the

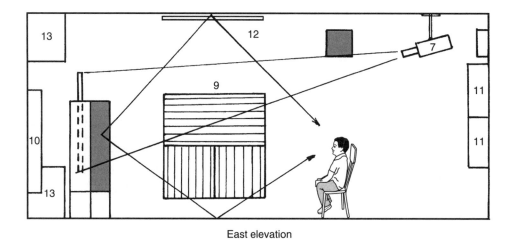

East elevation

West elevation

FIGURE 21-3   East and west elevation views of a home theater.

video screen (10) are made of glass-fiber sheet. Alternatively two Abffusors measuring 2 ft × 2 ft × 4 in can be used in each location; in this case, the upper unit is oriented for horizontal diffusion, and the lower unit is oriented for vertical diffusion. The diffuser behind the video screen would be identical to the two side-wall diffusers. For clarity, this diffuser application is shown in the figure.

In addition, bass traps (13) are mounted behind the video screen, all at one end of the room, one in each of the four tricorners. The location of these bass absorbers in the corners places them where all modes terminate; the corners are thus the most effective position. The bass traps suggested for this room are trapezoidal units that are designed for corner mounting (see Chap. 19). The face of the unit measures 2 × 2 ft. These units use internal glass-fiber sheets; if desired, the cavities between the glass-fiber sheets can be loosely stuffed with glass-fiber batt. Should the room be judged too "boomy," more bass absorbers should be mounted between the existing ones. Alternatively, Modex Corner absorbers (manufactured by RPG Diffusor Systems, Inc.) can be used for this application. This bass absorber (Fig. 16-11) can be designed for peak absorption of 40, 63, or 80 Hz; the latter two approximately coincide with the first-order 70.6-Hz axial mode associated with the height of the room. It is less effective for the modes associated with the length (30.4 Hz) and width (44.1 Hz) of the room, but it still provides absorption at these frequencies.

The north and south elevation sketches are shown in Fig. 21-4. The south elevation shows the location of four absorptive or diffusive panels (11). As noted, alternatively, four Abffusor panels can be used, oriented for vertical and horizontal diffusion. Also in the south elevation sketch, a possible location for the subwoofer (4) is indicated on the floor in the southeast corner of the room. In practice, the subwoofer should be positioned by ear as described below.

As shown in the north elevation sketch, both main front loudspeakers may require bases to bring them up to a desirable height. These should be simple 3/4-in plywood boxes, painted black and filled with sand to deaden the cavity resonance. There will be numerous pieces of electronic equipment associated with this home theater. None of this equipment should be placed in front of the listener, where it would produce undesirable early reflections. A short rack located behind the sofa is probably the best location. In that position it can be reached over the back of the sofa or a control panel could be located at the right end of the sofa as shown in Fig. 21-1. However, if any component has a cooling fan, it should be located away from the listener.

Thus far, in this design, attention has been given to the front main loudspeakers and subwoofer. Potential reflections from the center-channel loudspeaker should fall approximately in the same surface locations as those from the front main loudspeakers; however, this should be checked and the treatment panels moved or enlarged to cover these reflections. The surround speakers are primarily used to reproduce ambience and surround effects. This is particularly the case when dipole loudspeakers are used (as in surround movie playback). The diffusing panels on the rear wall and other treatment will assist in creating a satisfactory sound field. When front-firing surround speakers (as in surround music playback) are used, more care should be taken to ensure that no unwanted specular reflections are created. If needed, as described above, absorptive or diffusing panels should be placed as needed.

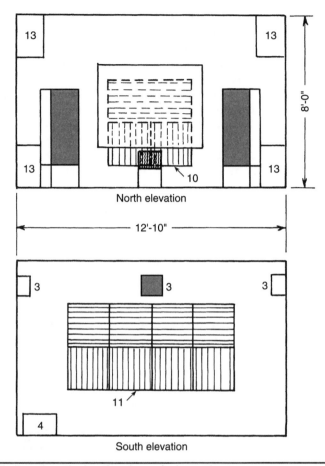

North elevation

12'-10"

South elevation

Figure 21-4   North and south elevation views of a home theater.

The Listening Environment

What can be expected in terms of the sound quality of the listening room? If early reflections from the front loudspeakers are well attenuated, their perceived frequency response will be nominally flat, and the sound field imaging will be sharp and definite. The location of the instruments in a musical group will be clear and the sound stage will be three-dimensional. Some degree of lateral reflections will provide a sense of spaciousness.

The center loudspeaker duplicates the function of that in a commercial movie theater. It reproduces and spatially anchors dialogue during movie playback, makes speech more intelligible, and helps enlarge the sweet spot for central images. The surround loudspeakers provide ambience that expands the sound field, and listeners will feel immersed in the music. The subwoofer reproduces the low-frequency components in music or a movie soundtrack. Movie sound effects such as explosions are perhaps the greatest beneficiary of a powered subwoofer.

How important is the quality of room acoustics in a home theater? The answer is that acoustics is very important, but so is the quality of the loudspeakers in the playback system (as well as the quality of the entire system itself). First, even the best acoustics cannot compensate for poor-sounding loudspeakers. Conversely, poor acoustics can seriously degrade the sound quality of good loudspeakers. Thus acoustics and loudspeaker quality are bound together; for high-quality performance, both must be good.

Various loudspeaker configurations have been suggested in the literature; they all rely on left/right symmetry of the five or more loudspeakers (not the subwoofer), and this is an important criteria. Otherwise, the choice of loudspeaker configurations is a matter of taste, and also depends on particular types of loudspeakers used and room geometry and furniture layout. The most widely used configuration, described in *ITU-R Recommendation BS.775-2, Multichannel Stereophonic Sound System With and Without Accompanying Picture*, (2006) has been found to work well. It places five loudspeakers at angles of 0°, ±30°, and ±110° to ±120° all relative to the listening position, as shown in Fig. 21-5. The rear angle can be successfully varied somewhat, again, according to taste and the type of surround loudspeakers used. Most listeners prefer to place the surround loudspeakers above seated ear level, by perhaps 2 ft.

The surround-sound playback system should be positioned so the front loudspeakers are placed along a short wall in a rectangular room. Unless necessary, loudspeakers (except for the subwoofer) should not be placed in corners; it is useful to have corner space available for bass traps, if needed.

The placement of the subwoofer depends on the particular type used, the room acoustics, and the listener's taste. One simple experiment is to temporarily place the subwoofer in the listening chair; while playing representative movies and music with good bass tracks, the listener moves throughout the room (on hands and knees, so your ears are close to the eventual height of the subwoofer) to find the place where the bass

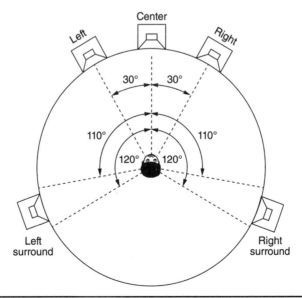

Figure 21-5 The ITU-R BS.775-2 standard for multichannel (5.1-channel) playback.

has a tight, clear, and solid sound. Then, the subwoofer is placed there. Usually, the best place for a subwoofer is in a corner; if this placement causes its output to be too loud, it is easy to attenuate it and adjust its crossover frequency. Multiple (perhaps two or four) subwoofers can provide a more uniform bass response over a larger listening area.

Reverberation Time

In this room design, absorption or diffusion has been used to reduce the effects of early reflections. When adding absorption, reverberation time must be checked to make sure the room is not too dead. Experience has shown that for a room of this size (1905 ft^3), a reverberation time of 0.3 to 0.5 sec at 500 Hz would be reasonable. A room with a reverberation time greater than 0.5 sec would risk poor speech intelligibility, and a room with reverberation time less than 0.3 sec would seem unnaturally dead and sound from the loudspeakers would be overly localized.

Because the effort would be imprecise, there is no point in initially designing for a specific reverberation time. A better approach would be to design for a reasonable range of reverberation times, build the room with reasonable treatment, then when construction is completed and the room is furnished, measure reverberation time, and listen to music and movies. With these evaluations, it can be determined whether the as-built reverberation time is suitable. Using this approach, it is advantageous to initially use slightly too-little absorption as it is easier to add absorption as needed, rather than remove it. Also, clearly, other problems such as flutter echo can be addressed after an initial evaluation. In this way, treatment can be finalized. As another approach, the treatment can be designed with some variability so that adjustments can be made easily after construction is completed. However, in most cases when treatment is designed with variability, its overall effect on reverberation time is usually quite limited.

Table 21-2 presents an example of a reverberation-time calculation for a home theater; because many of the treatment items are scratch-built, their absorption coefficients are estimated. Figure 21-6 plots the results of the calculation. The reverberation time ranges approximately from 0.3 to 0.6 sec. These times are within a reasonable range; however, the gypsum-board walls somewhat overabsorb the low frequencies. A concrete floor is assumed; a wood floor may add even more low-frequency absorption. If this is the case,

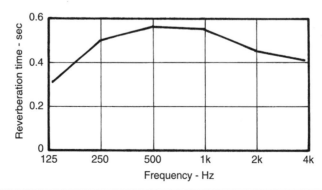

FIGURE 21-6 A plot of reverberation time for a home theater using the calculations in Table 21-2.

Description	Area, ft²	125 Hz		250 Hz		500 Hz		1 kHz		2 kHz		4 kHz	
		α	Sα	α	Sα	α	Sα	α	Sα	α	Sα	α	Sα
Drywall	740	0.29	214.6	0.10	74.0	0.05	37.0	0.04	29.6	0.07	51.8	0.09	66.6
Wall absorptive panel	80	0.82	65.6	0.90	72.0	1.00	80.0	1.00	80.0	1.00	80.0	1.00	80.0
Ceiling absorptive panel	16	0.66	10.6	0.95	15.2	1.00	16.0	1.00	16.0	1.00	16.0	1.00	16.0
Bass trap	16	0.40	6.4	0.20	3.2	0.10	1.6	0.05	0.8	0.03	0.5	0.02	0.3
Rug, 6 × 6 ft	36	0.02	0.7	0.06	2.2	0.14	5.0	0.37	13.3	0.60	21.6	0.65	23.4
Sofa	60	0.30	18.0	0.35	21.0	0.40	24.0	0.50	30.0	0.60	36.0	0.65	39.0
Total absorption, sabins			315.9		187.6		163.6		169.7		205.9		225.3
Reverberation time, sec			0.30		0.50		0.57		0.55		0.45		0.41

TABLE 21-2 Reverberation-Time Calculations for a Home Theater

it may be possible to reduce the number of bass traps. However, the bass response should be checked to ensure that the modal response does not become uncontrolled.

Very generally, within the range of reverberation times given above, a more live room is preferred for music playback, while a more dead room is better for movie playback and its associated speech intelligibility. The owner of the home theater should decide which function has priority, or whether a compromise is preferred. Some purpose-built home theaters are uncomfortably dead; this may be intentional on the part of the designer to make the room seem more acoustically solemn. More important than the designer's intent is the need for the room to sound "right" to the owners and listeners. Personal preference should never be ignored. As noted many times, room acoustics is both a science and an art.

APPENDIX

Selected Absorption Coefficients

Material	125 Hz	250 Hz	500 Hz	1 kHz	2 kHz	4 kHz	Reference
Building Materials							
Concrete block, coarse	0.36	0.44	0.31	0.29	0.39	0.25	1
Concrete block, painted	0.10	0.05	0.06	0.07	0.09	0.08	1
Glass, large heavy pane	0.18	0.06	0.04	0.03	0.02	0.02	1
Glass, window	0.35	0.25	0.18	0.12	0.07	0.04	1
Plaster, gypsum smooth, on brick	0.013	0.015	0.02	0.03	0.04	0.05	1
Plaster, gypsum smooth, on lath	0.14	0.10	0.06	0.05	0.04	0.03	1
Gypsum board 1/2-in on 2 × 4 studs, 16-in centers	0.29	0.10	0.05	0.04	0.07	0.09	1
Carpet, heavy on concrete	0.02	0.06	0.14	0.37	0.60	0.65	1
Carpet, heavy on 40-oz hairfelt	0.08	0.24	0.57	0.69	0.71	0.73	1
Carpet, heavy latex backing on foam or 40-oz hairfelt	0.08	0.27	0.39	0.34	0.48	0.63	1
Carpet, indoor-outdoor	0.01	0.05	0.10	0.20	0.45	0.65	2
Acoustical tile, 1/2-in average	0.07	0.21	0.66	0.75	0.62	0.49	
Acoustical tile, 3/4-in average	0.09	0.28	0.78	0.84	0.73	0.64	
Concrete floor	0.01	0.01	0.015	0.02	0.02	0.02	1
Floor: linoleum, asphalt tile, or cork on concrete	0.02	0.03	0.03	0.03	0.03	0.02	1
Floor: wood	0.15	0.11	0.10	0.07	0.06	0.07	1
Plywood panel, 3/8-in	0.28	0.22	0.17	0.09	0.10	0.11	1

(Continued)

Material	125 Hz	250 Hz	500 Hz	1 kHz	2 kHz	4 kHz	Reference
Polycylindrical Absorber							
Chord 45-in height 16-in empty	0.41	0.40	0.33	0.25	0.20	0.22	3
Chord 35-in height 12-in empty	0.37	0.35	0.32	0.28	0.22	0.22	3
Chord 28-in height 10-in empty	0.32	0.35	0.3	0.25	0.20	0.23	3
Chord 20-in height 8-in empty	0.25	0.30	0.33	0.22	0.20	0.21	3
Chord 20-in height 8-in filled	0.30	0.42	0.35	0.23	0.19	0.2	3
Perforated Panel							
5/32-in thick, 4-in depth, 2-in glass fiber							
Perforation 0.18%	0.40	0.70	0.30	0.12	0.10	0.05	3
Perforation 0.79%	0.40	0.84	0.40	0.16	0.14	0.12	3
Perforation 1.4%	0.25	0.96	0.66	0.26	0.16	0.10	3
Perforation 8.7%	0.27	0.84	0.96	0.36	0.32	0.26	3
5/32-in thick, 8-in depth, 4-in glass fiber							
Perforation 0.18%	0.80	0.58	0.27	0.14	0.12	0.10	3
Perforation 0.79%	0.98	0.88	0.52	0.21	0.16	0.14	3
Perforation 1.4%	0.78	0.98	0.68	0.27	0.16	0.12	3
Perforation 8.7%	0.78	0.98	0.95	0.53	0.32	0.27	3
7-in airspace, 1-in mineral fiber, 9–10 lb/ft^3 density, 1/4-in cover							
Wideband, 25% perforation	0.67	1.09	0.98	0.93	0.98	0.96	3
Midpeak, 5% perforation	0.60	0.98	0.82	0.90	0.49	0.30	3
Lowpeak, 0.5% perforation	0.74	0.53	0.40	0.30	0.14	0.16	3
2-in airspace, filled with mineral fiber, 9–10 lb/ft^3 density							
Perforation 0.5%	0.48	0.78	0.60	0.38	0.32	0.16	3
Key to References							
See references starting on page 337 for complete citations.							
1. Hedeen, 1980							
2. Seikman, 1969							
3. Mankovsky, 1971							

Glossary

A-weighting A frequency-response adjustment of a sound-level meter that makes its reading approximately conform to human hearing response.

absorption In acoustics, the changing of sound energy to heat.

absorption coefficient The fraction of sound energy absorbed at a given surface. It has a theoretical value between 0 and 1 and varies with frequency and angle of incidence.

acoustics The science that deals with the production, control, transmission, reception, and effects of sound.

active sound absorber A resonant sound-absorbing structure.

AES Audio Engineering Society

ambience The distinctive acoustical characteristics of a given space.

amplitude The instantaneous magnitude of an oscillating quantity such as sound pressure. The peak value is the maximum value.

amplitude distortion Distortion of the waveshape of the signal.

analog A signal whose frequency and level vary continuously in direct relationship to the original electrical or acoustical signal.

anechoic Without echo.

anechoic chamber A room designed to suppress internal sound reflections; used for acoustical measurements and testing.

arrival gap The time between the arrival of the direct signal and reflections. *See also* initial time-delay gap.

articulation A quantitative measure of the intelligibility of speech; the percentage of speech items correctly perceived and recorded.

artificial reverberation Reverberation generated by electrical or acoustical means to simulate the acoustics of spaces such as concert halls, etc.

ASA Acoustical Society of America.

ASHRAE American Society of Heating, Refrigerating, and Air-Conditioning Engineers.

attack The beginning of a sound; the initial transient of a musical note.

attenuate   To reduce the level of an acoustical or electrical signal.

attenuator   A device, usually a variable resistance, used to control the level of an electrical signal.

audio frequency   An acoustical or electrical signal of a frequency that falls within the audible range of the human ear, usually taken as 20 Hz to 20 kHz.

audio spectrum   The range of frequencies audible to the human ear.

auditory area   The sensory area lying between the threshold of hearing and the threshold of feeling or pain.

auditory system   The human hearing system made up of the external ear, the middle ear, the inner ear, the nerve pathways and the brain.

aural   Having to do with the auditory mechanism.

axial mode   A room mode produced by reflections from one pair of parallel surfaces of a room.

baffle   A movable barrier used in recording studios to improve separation of signals from different sources. Also, the surface or board upon which a loudspeaker is mounted.

bandpass filter   A filter that attenuates signals both below and above the desired passband.

bandwidth   The frequency range passed by a given device or structure. Commonly measured as the width between –3-dB points.

basilar membrane   A membrane inside the cochlea that vibrates in response to sound, exciting the hair cells.

bass   The lower range of audible frequencies.

bass boost   An increase in level of the lower range of frequencies, achieved by electrical circuits or acoustical reinforcement.

bass trap   A structure designed to absorb low-frequency sound energy.

beats   Periodic fluctuations resulting from superimposing signals of slightly different frequency.

binaural   Listening with two ears or recording with two microphones to simulate hearing.

boomy   A colloquial expression for excessive bass response.

byte   A term used in digital systems. One byte is equal to eight bits of data. A bit is the elemental "low" or "high" state of a binary system.

capacitor   An electrical component that passes alternating current but blocks direct current. Also called a condenser, it is capable of storing electrical energy.

clipping   The amplitude of the peaks of an electrical signal limited by electronic circuits or by overloading an electronic device. It is a distortion of the signal.

cochlea   The portion of the inner ear that changes mechanical vibrations to electrical signals. It is essentially the frequency-analyzing portion of the auditory system.

coincidence effect   Sound energy falling on a wall having a frequency coincident with the natural period of the wall sustains the wall vibration. This results in a decrease in the transmission loss for the wall near that frequency.

coloration   A distortion of an audio signal that is detectable by the ear.

comb filter   A distortion produced by combining an electrical or acoustical signal with a delayed replica of itself. The result is constructive and destructive interference that introduces peaks and nulls into the frequency response. When plotted on a linear frequency scale, the response resembles a comb, hence the name.

compression   Reducing the dynamic range of a signal with electrical circuits that reduce the level of loud passages.

condenser   *See* capacitor.

correlogram   A graph showing the correlation of one signal with another.

critical band   In human hearing, only those frequency components within a narrow band, called the critical band, will mask a given tone. Critical bandwidth varies with frequency, but is usually between 1/6 and 1/3 octave wide.

crossover frequency   In a loudspeaker with multiple radiators, the crossover frequency is the −3-dB point of the network dividing the signal energy.

crosstalk   The signal of one channel or circuit interfering with another.

dB   *See* decibel.

dBA   A sound-level meter reading with an A-weighting network simulating the human-ear response at a loudness level of 40 phons.

dBB   A sound-level meter reading with a B-weighting network simulating the human-ear response at a loudness level of 70 phons.

dBC   A sound-level meter reading with a C-weighting network simulating the human-ear response at a loudness level of 100 phons.

dBZ   A sound-level meter reading with a zero weighting used to describe a flat frequency response from 10 Hz to 20 kHz ± 1.5 dB.

decade   Ten times any quantity or frequency range. The range of the human ear is about three decades.

decay rate   A measure of the decay of acoustical signals, expressed as a slope in dB/second.

decibel   The human ear responds logarithmically and it is convenient to deal in logarithmic units in audio systems. The bel is the logarithm of the ratio of two powers, and the decibel is one tenth of a bel. Abbreviated dB.

delay line   A device or algorithm employed to delay one signal with respect to another.

diaphragm   Any surface that vibrates in response to sound or is vibrated to emit sound, such as in microphones and loudspeakers. Also applied to wall and floor surfaces vibrating in response to sound.

dielectric   An insulating material. The material between the plates of a capacitor.

diffraction   The spatial distortion of a wavefront caused by the presence of an obstacle in the sound field.

diffuser   A device for the diffusion of sound; for example, through reflection phase-grating means.

diffusion   The process of diffusing or scattering of sound.

diffusion coefficient   The ratio of scattered intensity at 45° to the specular intensity.

digital audio   A numerical presentation of an audio signal. Pertaining to the application of digital techniques to audio tasks.

distance double   In pure spherical divergence of sound from a point source in free space, the sound-pressure level decreases 6 dB for each doubling of the distance.

distortion   Any change in waveform or harmonic content of an original signal as it passes through a device. The result of nonlinearity within the device.

distortion, harmonic   The change in the harmonic content of a signal when it passes through a nonlinear device.

dynamic range   All audio systems are limited by inherent noise at low levels and by overload distortion at high levels. The usable range between the two extremes is the dynamic range of the system. Expressed in dB.

dyne   The force that will accelerate a one-gram mass at the rate of 1 cm/sec. The old standard reference level for sound pressure was 0.0002 dyne/cm^2. The same level today is expressed as 20 micropascals.

ear canal   The external auditory meatus; the canal between the pinna and the eardrum.

eardrum   The tympanic membrane located at the end of the ear canal that is attached to the ossicles of the middle ear.

early sound   Direct and reflected components that arrive at the ear from a source during the first 50 msec or so. Such components are replicas of the original sound and arrive at different times producing comb-filter distortion. Also the basis of desirable effects such as spaciousness and defining of the stereo image.

echo   A delayed sound that is perceived by the ear as a discrete sound image.

echogram   A record of the early reverberatory decay of sound in a room.

EFC   Energy-frequency curve.

EFTC   Energy-frequency-time curve.

ensemble   Ability of musicians to hear each other to perform properly. Diffusing elements surrounding the stage area contribute to ensemble.

equalization   The process of adjusting the frequency response of a device or system to achieve a flat or other desired response.

equalizer   A device for adjusting the frequency response of a device or system.

equal loudness contour   A contour representing a constant loudness for all audible frequencies. The contour having a sound-pressure level of 40 dB at 1000 Hz is arbitrarily defined as the 40-phon contour.

ETC   Energy-time curve.

Eustachian tube   The tube running from the middle ear into the pharynx; it equalizes middle-ear atmospheric pressure.

external meatus   The ear canal terminated by the eardrum.

feedback, acoustic   Unwanted interaction between the output and the input of an acoustical system, for example, between the loudspeaker and the microphone of a system.

FFT   Fast Fourier transform. An iterative algorithm that computes the Fourier transform in a short time.

fidelity   As applied to sound quality, the faithfulness of a signal to the original.

filter, bandpass   A filter that passes all energy between a low-frequency cutoff frequency and a high-frequency cutoff frequency.

filter, high-pass   A filter that passes all energy above a cutoff frequency.

filter, low-pass   A filter that passes all energy below a cutoff frequency.

flanking sound   Sound traveling by circuitous paths which reduces the effectiveness of an insulating barrier.

floating floor   A massive floor that is resiliently connected to the structure for the purpose of increasing transmission loss.

flutter echo   A repetitive echo set up by parallel reflecting surfaces.

Fourier analysis   Application of the Fourier transform to a signal to determine its spectrum.

fractal   A property or shape which is repeated at progressively smaller scales. The property of self-similarity. Fractals applied to quadratic-residue diffusers result in extended range.

frequency   The measure of the rapidity of alternations of a periodic signal, expressed in hertz.

frequency response   The changes of the sensitivity of a circuit or device with frequency.

FTC   Frequency-time curve.

fundamental   The basic pitch or frequency of a musical note or a harmonic series of frequencies.

fusion zone   All reflections arriving at the observer's ears within 20 to 40 msec of the direct sound are integrated, or fused together, with a resulting apparent increase in level and a change of character. *See also* precedence effect.

gain   The increase in power level of a signal.

graphic-level recorder   A device for recording signal level vs. time. The level vs. angle can also be recorded for directivity patterns.

grating, diffraction   An optical grating consisting of minute parallel lines used to break down light into its component colors.

grating, reflection phase   An acoustical diffraction grating used to diffuse sound.

Haas effect   *See* precedence effect.

hair cell The sensory elements of the cochlea that transduce the mechanical vibrations of the basilar membrane to nerve impulses that are sent to the brain.

harmonic distortion *See* distortion, harmonic.

harmonics Integral multiples of a fundamental frequency. The first harmonic is the fundamental, and the second is twice the frequency of the fundamental, and so on.

hearing loss The loss of sensitivity of the auditory system measured in decibels below a standard level. Some hearing loss is age-related; some is related to exposure to high-level sound.

Helmholtz resonator A reactive, tuned sound absorber. A bottle is an example of such a resonator. Often made by placing a perforated cover or slats over a cavity.

henry The unit of inductance.

hertz The unit of frequency, abbreviated Hz. Cycles per second. The frequency of an electrical signal or sound wave.

high-pass filter *See* filter, high-pass.

HVAC Heating, ventilating, and air-conditioning.

IEEE Institute of Electrical and Electronics Engineers.

Impact Insulation Class (IIC) A single value rating used to quantify impact noise in a floor/ceiling.

impedance matching Maximum power is transferred from one circuit to another when the output impedance of one is matched to the input impedance of the other. Maximum power transfer may be less important than low noise or voltage gain in many electronic circuits.

impulse A very short, transient electrical or acoustical signal.

in phase Two periodic waves reaching peaks and going through zero at the same time are said to be "in phase."

inductance An electrical characteristic of circuits, especially of coils, that introduces inertial lag because of the presence of a magnetic field. Measured in henrys.

initial time-delay gap The time gap between the arrival of the direct sound and the first sound reflected from the surfaces of the room. *See* arrival gap.

insulation As referred to sound barriers, insulation refers to the sound transmission loss of a particular wall, floor/ceiling, etc.

intensity Acoustic intensity is sound energy flux per unit area. The average rate of sound energy transmitted through a unit area normal to the direction of sound transmission.

interference The combining of two or more signals results in an interaction called interference. This can be constructive or destructive. Another use of the term refers to undesired, intrusive signals.

intermodulation distortion Distortion produced by the interaction of two or more signals. The distortion components are not harmonically related to the original signals.

inverse distance law In a free field, the sound-pressure level decreases by 6 dB as the distance from the source is doubled.

isolation   Refers to the isolation of an entire studio or room from outside noise.

ITD   Initial time delay, as in an ITD gap.

JAES   Journal of the Audio Engineering Society.

JASA   Journal of the Acoustical Society of America.

kHz   1000 Hz.

law of the first wavefront   The first wavefront falling on the ear determines the perceived direction of the sound.

level   A sound-pressure level expressed in decibel means that it is calculated with respect to the standard reference level of 20 micropascals. The word "level" associates that figure with the appropriate standard reference level.

linear   A device or circuit is linear if a signal passing through is not distorted.

live end-dead end (LEDE)   An acoustical treatment plan for rooms, in which the front end is highly absorbent and the rear end is reflective and diffusive.

logarithm   An exponent of 10 in the common logarithms to the base 10. For example, 10 to the exponent 2 = 100; the log of 100 = 2.

loudness   A subjective term for the sensation of the magnitude of sound.

loudspeaker   An electroacoustical transducer that changes electrical energy to acoustical energy.

masking   The amount, or the process, by which the threshold of audibility for one (masked) sound is raised by the presence of another (masking) sound.

mass-air-mass resonance   A resonating system composed, for example, of the mass of two spaced glass panes and the air between them. There is usually a dip in the transmission-loss curve at the frequency at which this system is resonant.

mean free path   For sound waves in an enclosure, the average distance traveled between successive reflections.

microphone   An electroacoustical transducer that changes acoustical energy to electrical energy.

middle ear   The cavity between the eardrum and the cochlea in which the ossicles connect the eardrum and the oval window.

mixing console   A device used for processing audio signals from many sources and combing them.

mode   A room resonance. There are axial, tangential, and oblique modes.

monaural   *See* monophonic.

monitor   A loudspeaker used in the control room of a recording studio.

multitrack   A system of recording multiple tracks. The signals recorded on the various tracks are then mixed down to obtain the final recording.

NAB   National Association of Broadcasters.

noise   Interference of an electrical or acoustical nature. Noise can be a desirable signal in some applications.

noise criteria   Standard spectrum curves by which a given measured noise may be described by a single NC or NCB number.

nonlinear   A device or circuit is nonlinear if a signal passing through it is distorted.

normal mode   A room resonance. *See* mode.

null   A low or minimum point on a waveform. A minimum pressure region in a room.

oblique mode   A room mode produced by reflections from all six surfaces of a rectangular room.

octave   The interval between two frequencies having a ratio of 2:1.

oscilloscope   An indicating instrument used to display waveforms, for example, on a time axis.

ossicles   A linkage of three tiny bones providing the mechanical coupling between the eardrum and the oval window of the cochlea consisting of the hammer, anvil, and stirrup.

out of phase   The offset in time of two related signals.

oval window   A tiny membranous window on the cochlea to which the footplate of the stirrup ossicle is attached. The sound from the eardrum is transmitted to the fluid of the inner ear through the oval window.

overtone   A component of a complex tone having a frequency higher than the fundamental frequency.

panel absorber   A panel mounted with an enclosed airspace that vibrates and absorbs sound energy.

partial   One of a group of frequencies, not necessarily harmonically related to the fundamental, that appears in a complex tone.

passive absorber   A sound absorber that dissipates sound energy as heat.

perforated absorber   A panel absorber with perforated holes in the panel and an enclosed airspace functioning as a Helmholtz absorber.

phase   The time relationship between two signals.

phon   The unit of loudness level of a tone.

pink noise   A noise signal whose spectrum level decreases at a 3 dB/octave rate. This gives the noise equal energy per octave.

pinna   The exterior ear.

pitch   A subjective term for the perceived frequency of a tone.

plenum   An absorbent-lined building cavity through which conditioned air is routed to reduce noise.

polar pattern   A graph showing the directional characteristics of a microphone or a loud-speaker.

polarity   The relative position of the high (+) and the low (–) signal leads in an audio system.

precedence effect   Delayed sounds are integrated by the auditory apparatus if they fall on the ear within 20 to 40 msec of the direct sound. The level of the delayed components contributes to the apparent level of the sound, and sound is localized at the first-arriving source. Also called the Haas effect. *See also* fusion zone.

primitive root   The root of a prime number.

psychoacoustics   The study of the interaction of the auditory system and acoustics.

pure tone   A tone with no harmonics. All energy is concentrated at a single frequency.

Q factor   Quality factor. A measure of the losses in a resonance system. The sharper the resonance curve, the higher the Q.

quadratic-residue sequence   A mathematical expression used in the design of diffusers to determine well depth.

random noise   A noise signal which has constantly shifting amplitude, phase, and frequency, and a uniform spectral distribution of energy.

ray   At higher audio frequencies sound can be considered as rays traveling in straight lines in a direction normal to the wavefront.

reactive absorber   A sound absorber, such as a Helmholtz resonator, which utilizes the effects of mass and compliance as well as resistance.

reactive silencer   A silencer in air-conditioning systems that uses reflection losses to provide attenuation.

reflection   For surfaces large compared to the wavelength of impinging sound, sound is reflected much as light is reflected, as a specular reflection with the angle of reflection equal to the angle of incidence.

reflection-phase grating   A diffuser of sound energy using the principle of the diffraction grating.

refraction   The bending of sound waves traveling through layered media with different propagation velocities.

resistance   That quality of electrical or acoustical circuits that results in dissipation of energy through heat.

resonance   A natural periodicity, or the reinforcement associated with this periodicity.

resonator silencer   An air-conditioning silencer employing tuned stubs and their resonating effect to provide attenuation.

reverberation   The decay of sound in an enclosure after the source has stopped. Caused by multiple reflections from the room boundaries.

reverberation time   The time required for the sound in an enclosure to decrease a certain amount, usually by 60 dB. The latter is abbreviated as RT_{60}.

ringing   The tendency of high-Q electrical circuits and acoustical devices to oscillate (or ring) when excited by a suddenly applied signal.

round window   The tiny membrane of the cochlea that opens into the middle ear cavity serving as a pressure release for the cochlea fluid.

RT$_{60}$ Reverberation time. Time for sound to decay by 60 dB. It can be measured at different frequencies, but 500 Hz is often used by default.

sabin   A unit of sound absorption. One square foot of open window has an absorption of 1 sabin.

semicircular canals   The three sensory organs for balance that are a part of the cochlear structure.

sequence, maximum length   A measurement method using a pseudo-random binary sequence to excite a system and/or room with a test signal resembling wideband white noise. Also a mathematical sequence used in determining the well depth of diffusers.

sequence, primitive root   A mathematical sequence used in determining the well depth of diffusers.

sequence, quadratic residue   A mathematical sequence used in determining the well depth of diffusers.

signal-to-noise ratio   The difference measured in decibels between the nominal or maximum operating level and the noise floor.

sine wave   A periodic wave related to simple harmonic motion.

slap-back   A discrete reflection from a nearby surface.

sone   The unit of measurement for subjective loudness.

sound absorption coefficient   The practical unit between 0 and 1 (nominally) expressing the absorbing efficiency of a material. Coefficient values are determined experimentally.

sound-level meter   A microphone-amplifier-meter device calibrated to read sound-pressure level above the reference level of 20 micropascals.

sound-power level   A power expressed in decibel above the standard reference level of 1 picowatt.

sound spectrograph   An instrument that displays the time, level, and frequency of a signal.

spectrum   The distribution of the energy of a signal with respect to frequency.

spectrum analyzer   An instrument for measuring and/or recording the spectrum of a signal.

specularity   A term devised to express the efficiency of diffraction-grating types of diffusers.

specular reflection   A mirror-like reflection of sound (or light) from a surface with the angle of reflection equal to the angle of incidence.

spherical divergence   Sound diverges spherically from a point source in free space.

splaying   Walls are splayed when they are constructed somewhat "off-square," that is, a few degrees from the rectilinear form.

standing wave   A resonance condition in an enclosed space in which sound waves traveling in one direction interact with those traveling in the opposite direction, resulting in a stable condition.

STC   Sound Transmission Class. A single-number system of designating sound transmission loss of partitions, etc.

steady-state   A condition devoid of transient effects.

stereo   A stereophonic system of two channels.

superposition   Many sound waves may traverse the same point in space, the air molecules responding to the vector sum of the different waves.

tangential mode   A room mode produced by reflections from four of the six surfaces of a rectangular room.

threshold of feeling (pain)   The sound-pressure level that produces a perceived physical sensation in the ears at approximately 120 dB above the threshold of hearing.

threshold of hearing   The lowest sound level that can be perceived by the human auditory system. This is close to the standard reference level of sound pressure, 20 micropascals.

timbre   The quality of a sound related to its harmonic structure.

tone   A tone results in the auditory sensation of pitch.

tone burst   A short signal used in acoustical measurements making it possible to differentiate desired signals from spurious reflections.

tone control   An electrical circuit to allow adjustment of frequency response.

transducer   A device for changing electrical signals to acoustical or vice versa, such as a microphone or loudspeaker.

transient   A short-lived aspect of a signal, such as the attack of a percussive instrument.

treble   The higher frequencies of the audible spectrum.

volume   The colloquial equivalent of sound level.

watt   The unit of electrical or acoustical power.

wave   A regular variation of an electrical signal or acoustical pressure.

wavelength   The distance a sound wave travels in the time it takes to complete one cycle.

weighting   Adjustment of sound-level meter response to achieve a desired measurement.

white noise   Random noise having uniform distribution of energy with frequency.

References and Resources

References

Allison, R. F., "Influence of Listening Room on Loudspeaker Systems," *Audio*, 63:8, pp. 37–40, Aug. 1979.

American Society of Heating, Refrigerating, and Air-Conditioning Engineers, Inc., *ASHRAE Handbook—Systems*, ASHRAE, Inc., 1984.

Ando, Y., and P. Cariani, *Auditory and Visual Sensations*, Springer, 2009.

ASTM, *Standard Classification for Determination of Sound Transmission Class*, American Society for Testing and Materials, Standard E413-87 (R1994).

Bartel, T. W., "Effect of Absorber Geometry on Apparent Absorption Coefficients as Measured in a Reverberation Chamber," *Journal of the Acoustical Society of America*, 69:4, pp. 1065–1074, April 1981.

Beranek, L. L., *Music, Acoustics, and Architecture*, John Wiley & Sons, 1962.

Beranek, L. L., "Balanced Noise-Criterion (NCB) Curves," *Journal of the Acoustical Society of America*, 86:2, pp. 69–101, Aug. 1989.

Beranek, L. L., "Concert Hall Acoustics - 1992," *Journal of the Acoustical Society of America*, 92:1, pp. 1–39, July 1992.

Blazier, W., Jr. and R. B. DuPree, "Investigation of Low-Frequency Footfall Noise in Wood-Frame Multifamily Building Construction," *Journal of the Acoustical Society of America*, 96:3, pp. 1521–1532, Sept. 1994.

Bolt, R. H., "Note on Normal Frequency Statistics for Rectangular Rooms," *Journal of the Acoustical Society of America*, 18:1, pp. 130–133, 1946.

Bonello, O. J., "A New Criterion for the Distribution of Normal Room Modes," *Journal of the Audio Engineering Society*, 29:9, pp. 597–606, Sept. 1981; erratum, ibid., p. 905, Dec. 1981.

Cavanaugh, W. J., and J. A. Wilkes, eds., *Architectural Acoustics: Principles and Practice*, John Wiley & Sons, 1999.

Cervone, R. P., "Subjective and Objective Methods for Evaluating the Acoustical Quality of Buildings for Music," Master of Architecture Thesis, University of Florida, 1990.

Cops, A., H. Myncke, and G. Vermeir, "Insulation of Reverberant Sound through Double and Multilayered Glass Constructions," *Acustica*, 23, pp. 257–265, 1975.

Cowan, J. P., *Architectural Acoustics: Design Guide*, McGraw-Hill, 2000.

Cowan, J. P., *Handbook of Environmental Acoustics*, Van Nostrand Reinhold, 1994.

Cowan, J. P., Personal Noise Monitoring Data, 1992.

Cox, T. J., and P. D'Antonio, *Acoustic Absorbers and Diffusers: Theory, Design and Application*, Spon Press, 2004.

Cox, T. J., P. D'Antonio, and M. R. Avis, "Room Sizing and Optimization at Low Frequencies," *Journal of the Audio Engineering Society*, 52:6, June 2004.

D'Antonio, P., and J. H. Konnert, "The Reflection Phase Grating Diffusor: Design, Theory, and Application," *Journal of the Audio Engineering Society*, 32:4, pp. 228–238, April 1984.

D'Antonio, P., and J. H. Konnert, "The QRD Diffractal: A New 1- or 2-Dimensional Fractal Sound Diffusor," *Journal of the Audio Engineering Society*, 40:3, pp. 117–129, March 1992.

D'Antonio, P., J. H. Konnert, and R. E. Berger, "Control Room Design Utilizing a Reflection Free Zone and Reflection Phase Grating Acoustical Diffusors: A Case Study," *78th Convention of the Audio Engineering Society*, (presentation only), May 1985.

D'Antonio, P., C. Bilello, and D. Davis, "Optimizing Home Listening Rooms, Part 1," *85th Convention of the Audio Engineering Society*, preprint 2735, Nov. 1988.

Davis, D., and C. Davis, "The LEDE™ Concept for Control of Acoustic and Psychoacoustic Parameters in Recording Control Rooms," *Journal of the Audio Engineering Society*, 28:9, pp. 585–595, Sept. 1980.

EBU R22-1998, "Listening Conditions for the Assessment of Sound Programme Material," *Technical Recommendation*, European Broadcasting Union, 1998.

Egan, M. D., *Architectural Acoustics*, McGraw-Hill, 1988.

Egan, M. D., *Concepts in Architectural Acoustics*, McGraw-Hill, 1972.

Everest, F. A., and K. C. Pohlmann, *Master Handbook of Acoustics*, McGraw-Hill, 5th ed., 2009.

Flynn, D. R., et al., *Acoustical and Thermal Performance of Exterior Residential Walls, Doors, and Windows*, National Bureau of Standards, Technical Publication PB-246–716/5, Nov. 1975.

Gilford, C. L. S., *Acoustics for Radio and Television Studios*, Peter Peregrinus, Ltd., 1972.

Gilford, C. L. S., "The Acoustic Design of Talk Studios and Listening Rooms," 1959, reprinted in *Journal of the Audio Engineering Society*, 27:1/2, pp. 17–31, Jan/Feb. 1979.

Grantham, J. B., and T. B. Heebink, "Sound Attenuation Provided by Several Wood-Frame Floor/Ceiling Assemblies with Troweled Floor Toppings," *Journal of the Acoustical Society of America*, 54:2, pp. 353–360, 1973.

Green, D. W., and C. W. Sherry, "Sound Transmission Loss of Gypsum Wallboard Partitions," *Journal of the Acoustical Society of America*, Report #1: Unfilled Steel Stud Partitions, 71:1, pp. 90–96, Jan. 1982; Report #2: Steel Stud Partitions Having Cavities Filled with Fiber Batts, 71:4, pp. 902–907, April 1982; Report #3: 2 × 4 Wood Stud Partitions, 71:4, pp. 908–914, April 1982.

Harris, C. M., *Handbook of Noise Control*, McGraw-Hill, 1957.

Hedeen, R. A., *A Compendium of Materials for Noise Control*, DHEW (NIOSH) publication No. 80-116, U.S. Government Printing Office, 1980.

Heringa, P. H., "Comparison of the Quality for Music of Different Halls," 11th Intl. Congress on Acoustics, Paris, 7, July 1983.

Hirschorn, M., "Fiberglass & Noise Control—Is It a Safe Combination?" *Sound and Vibration*, 28:10, pp. 6–10, Oct. 1994.

ITU-R Recommendation BS. 775-2, "Multichannel Stereophonic Sound System With and Without Accompanying Picture," 2006.

Jackson, G. M., and H. G. Leventhall, "The Acoustics of Domestic Rooms," *Applied Acoustics*, 5, pp. 265–277, 1972.

Kuttruff, H., *Room Acoustics*, 5th ed., Spon Press, 2009.

Libby-Owens-Ford Company (LOF), *Sound Reduction Design Considerations for Construction Glass*, Libby-Owens-Ford Company.

Lockner, J. P. A., and J. F. Burger, "The Subjective Masking of Short Time-Delayed Echoes by their Primary Sounds and their Contribution to the Intelligibility of Speech," *Acustica*, 8, pp. 1–10, 1958.

Long, M., *Architectural Acoustics*, Elsevier Academic Press, 2006.

Louden, M. M., "Dimension Ratios of Rectangular Rooms with Good Distribution of Eigentones," *Acustica*, 24, pp. 101–104, 1971.

Mankovsky, V. S., *Acoustics of Studios and Auditoria*, Focal Press, Ltd., 1971.

Mehta, M., J. Johnson, and J. Rocafort, *Architectural Acoustics: Principles and Design*, Prentice-Hall, 1999.

Meyer, E., and G. R. Schodder, *On the Influence of Reflected Sound on Directional Localization and Loudness of Speech*, Nachr. Akad. Wiss., Göttingen, Physics, Klasse IIa, 6, pp. 31–42, 1952.

Morse, P., and R. H. Bolt, "Sound Waves in Rooms," *Reviews of Modern Physics*, 16:2, pp. 69–150, April 1944.

National Bureau of Standards (NBS), *Acoustical and Thermal Performance of Exterior Residential Walls, Doors, and Windows*, NBS Technical Publication PB-246–716, 1975.

Nimura, T., and K. Shibayama, "Effect of Splayed Walls of a Room on Steady-State Sound Transmission Characteristics," *Journal of the Acoustical Society of America*, 29:1, pp. 85–93, Jan. 1957.

Northwood, T. D., *Transmission Loss of Plasterboard Walls*, Building Research Note no. 66, Division of Building Research, National Research Council, 1968.

Olive, S. E., and F. R. Toole, "The Detection of Reflections in Typical Rooms," *Journal of the Audio Engineering Society*, 37:7/8, pp. 539–553, July/Aug. 1989.

Pelton, H. K., S. Wise, and W. Sims, "Active HVAC Noise Control Systems Provide Acoustical Comfort," *Sound and Vibration*, 28:7, pp. 6–13, July 1994.

Pisha, B., and C. Bilello, "Designing a Home Listening Room," *Audio*, pp. 48–58, Aug. 1987.

Quirt, J. D., "Sound Transmission Through Windows: I. Single and Double Glazing," *Journal of the Acoustical Society of America*, 72:3, pp. 834–844, Sept. 1982.

Quirt, J. D., "Sound Transmission through Windows: II. Double and Triple Glazing," *Journal of the Acoustical Society of America*, 74:2, pp. 834–844, Aug. 1983.

RPG Diffusor Systems, Inc., *The RPG Home Concert Hall*, RPG Diffusor Systems, Inc., 1990.

Ruzicka, J. E., "Fundamental Concepts of Vibration Control," *Sound and Vibration*, 5:7, pp. 16–23, July 1971.

Sanders, G. J., "Silencers: Their Design and Applications," *Sound and Vibration*, 2:2, pp. 6–13, Feb. 1968.

Schroeder, M. R., "Diffuse Sound Reflections by Maximum-Length Sequence," *Journal of the Acoustical Society of America*, 57:1, pp. 149–150, Jan. 1975.

Schroeder, M. R., "Binaural Dissimilarity and Optimum Ceilings for Concert Halls: More Lateral Sound Diffusion," *Journal of the Acoustical Society of America*, 65:4, pp. 958–963, April 1979.

Schroeder, M. R., *Number Theory in Science and Communication*, 2nd. ed. Springer-Verlag, 1988.

Sepmeyer, L. W., "Computed Frequency and Angular Distribution of the Normal Modes of Vibration of Rectangular Rooms," *Journal of the Acoustical Society of America*, 37:3, pp. 413–423, March 1965.

Shea, M., and F. A. Everest, *How to Build a Small Budget Recording Studio from Scratch*, 4th ed., McGraw-Hill, 2012.

Siebein, G. W., "A Method to Evaluate the Acoustical Consequences of Conceptual Decisions in the Studio Design Process," *Proceedings of the 1986 ACSA Technology Conference*, 1988.

Siebein, G. W., and M. A. Gold, "The Concert Hall of the 21st Century: Historic Precedent and Virtual Reality. Architecture: Material and Imagined," *Proceedings of the 85th ACSA Annual Meeting*, pp. 52–61, 1997.

Siekman, W., "Outdoor Acoustical Treatment: Grass and Trees," *Journal of the Acoustical Society of America*, 46:4A, pp. 863–864, Oct. 1969.

Toole, F. E., *Sound Reproduction: Loudspeakers and Rooms*, Elsevier Focal Press, 2008.

Volkmann, J. E., "Polycylindrical Diffusers in Room Acoustic Design," *Journal of the Acoustical Society of America*, 13:3, pp. 234–243, Jan. 1941.

van Nieuwland, J. M., and C. Weber, "Eigenmodes in Non-Rectangular Reverberation Rooms," *Noise Control Engineering*, 13:3, pp. 112–121, Nov./Dec. 1979.

Walker, R., "Optimum Dimension Ratios for Small Rooms," *100th Convention of the Audio Engineering Society*, Preprint 4191, May 1996.

Warnock, A., *How to Reduce Noise Transmission Between Apartments*, Building Research Note No. 44. Division of Building Research, National Research Council, 1983.

Welti, T., and A. Devantier, "Low-Frequency Optimization Using Multiple Subwoofers," *Journal of the Audio Engineering Society*, 54:5, pp. 347–364, May 2006.

Resources

Acoustic Sciences Corporation (ASC)
4275 W. 5th Ave.
Eugene, OR 97402
(800) 272-8823
acousticsciences.com

Acoustical Surfaces
123 Columbia Ct. North, Suite 201
Chaska, MN 55318
(800) 854-2948
acousticalsurfaces.com

Auralex Acoustics, Inc.
9955 Westpoint Dr., Suite 101
Indianapolis, IN 46256
(800) 959-3342
auralex.com

BRD Noise and Vibration Control
112 Fairview Ave.
Wind Gap, PA 18091
(610) 863-6300
brd-nonoise.com

Brejtfus
410 S. Madison Dr.

Tempe, AZ 85281
(480) 731-9899
550 Front St. #2802
San Diego, CA 92101
(619) 813-6972
brejtfus.com

Brüel & Kjaer
World Headquarters
Skodsborgvej 307
DK-2850
Naerum, Denmark
+45 7741 2000
bksv.com

Duracote Corp.
350 N. Diamond St.
Ravenna, OH 44266
(800) 321-2252
duracote.com

E.A.R. Specialty Composites
7911 Zionsville Rd.
Indianapolis, IN 46268
(877) 327-4332
earsc.com

Eckel Noise Control Technologies
155 Fawcett St.
Cambridge, MA 02138
(617) 491-3221
eckelacoustic.com

Glass Association of North America
800 SW Jackson St., Suite 1500
Topeka, KS 66612
(785) 271-0208
glasswebsite.com

Industrial Noise Control, Inc.
401 Airport Rd.
North Aurora, IL 60542
(800) 954-1998
industrialnoisecontrol.com

Johns Manville Corp.
P.O. Box 5108
Denver, CO 80217

(303) 978-2000
jm.com

Kinetics Noise Control, Inc.
6300 Irelan Place
Dublin, OH 43017
(877) 457-2695
kineticsnoise.com

Larson-Davis, Inc.
1681 West 820 North
Provo, UT 84601
(801) 375-0177
larsondavis.com

Linear Products Corp.
P.O. Box 902
Cranford, NJ 07016
(908) 272-2211
Linearproducts.net

Mason Industries
350 Rabro Dr.
Hauppauge, NY 11788
(631) 348-0282
mason-industries.com

Metal Form Manufacturing, Inc.
5960 W. Washington St.
Phoenix, AZ 85043
(602) 233-1211
mfmca.com

Overly Manufacturing Company
P.O. Box 70
Greensburg, PA 15601
(800) 979-7300
overlymanufacturing.com

Owens-Corning
Fiberglas Tower
Toledo, OH 43659
(800) 438-7465
owenscorning.com

Peer, Inc. (makers of Almute panels)
2300 Norman Dr.
Waukegan, IL 60085

(847) 785-2900
peerinc.com

Pinta Acoustic, Inc. (makers of Sonex)
2601 49th Ave. N., Suite 400
Minneapolis, MN 55430
(800) 662-0032
pinta-acoustic.com

Primacoustic
1588 Kebet Way
Port Coquitlam, British Columbia
Canada V3C 5M5
(604) 942-1001
primacoustic.com

Proudfoot Company, Inc.
588 Pepper St.
Monroe, CT 06468
(800) 445-0034
theproudfootcompany.com

Quest Technologies
1060 Corporate Center Dr.
Oconomowoc, WI 53066
(800) 245-0779
questtechnologies.com

RPG Diffusor Systems, Inc.
651-C Commerce Dr.
Upper Marlboro, MD 20774
(301) 249-0044
rpginc.com

Tectum, Inc.
P.O. Box 3002
Newark, OH 43058
(888) 977-9691
tectum.com

UFP Technologies
707 Umatilla St.
Denver, CO 80204
(800) 372-3172
enmurray.com

Index

━ H ━

━ I ━